Mechanisms and Management of Senescence

Edited by Hassan M. Heshmati

Published in London, United Kingdom

Mechanisms and Management of Senescence
http://dx.doi.org/10.5772/intechopen.95366
Edited by Hassan M. Heshmati

Contributors
Hassan M. Heshmati, Heriberto Moran, Shanaz A. Ghandhi, Naoko Shimada, Karen Hubbard, Tapash Jay Sarkar, Maiko Hermsmeier, Jessica L. Ross, G. Scott Herron, Danielle Wang, Haitao Wang, Jelena Milić, Raghad Alshadidi

First published in London, United Kingdom, 2022 by IntechOpen
IntechOpen is the global imprint of INTECHOPEN LIMITED, registered in England and Wales, registration number: 11086078, 5 Princes Gate Court, London, SW7 2QJ, United Kingdom
Printed in Croatia

British Library Cataloguing-in-Publication Data
A catalogue record for this book is available from the British Library

Additional hard and PDF copies can be obtained from orders@intechopen.com

Mechanisms and Management of Senescence
Edited by Hassan M. Heshmati
p. cm.

This title is part of the Physiology Book Series, Volume 15
Topic: Cell Physiology
Series Editor: Tomasz Brzozowski
Topic Editor: Angel Catala

Print ISBN 978-1-83969-050-1
Online ISBN 978-1-83969-051-8
eBook (PDF) ISBN 978-1-83969-052-5
ISSN 2631-8261

For EU product safety concerns:
IN TECH d.o.o., Prolaz Marije Krucifikse Kozulić 3, 51000 Rijeka, Croatia,
info@intechopen.com or visit our website at intechopen.com.

IntechOpen Book Series

Physiology

Volume 15

Aims and Scope of the Series

Modern physiology requires a comprehensive understanding of the integration of tissues and organs throughout the mammalian body, including the cooperation between structure and function at the cellular and molecular levels governed by gene and protein expression. While a daunting task, learning is facilitated by identifying common and effective signaling pathways mediated by a variety of factors employed by nature to preserve and sustain homeostatic life. As a leading example, the cellular interaction between intracellular concentration of Ca+2 increases, and changes in plasma membrane potential is integral for coordinating blood flow, governing the exocytosis of neurotransmitters, and modulating gene expression and cell effector secretory functions. Furthermore, in this manner, understanding the systemic interaction between the cardiovascular and nervous systems has become more important than ever as human populations' life prolongation, aging and mechanisms of cellular oxidative signaling are utilised for sustaining life. Altogether, physiological research enables our identification of distinct and precise points of transition from health to the development of multimorbidity throughout the inevitable aging disorders (e.g., diabetes, hypertension, chronic kidney disease, heart failure, peptic ulcer, inflammatory bowel disease, age-related macular degeneration, cancer). With consideration of all organ systems (e.g., brain, heart, lung, gut, skeletal and smooth muscle, liver, pancreas, kidney, eye) and the interactions thereof, this Physiology Series will address the goals of resolving (1) Aging physiology and chronic disease progression (2) Examination of key cellular pathways as they relate to calcium, oxidative stress, and electrical signaling, and (3) How changes in plasma membrane produced by lipid peroxidation products can affect aging physiology, covering new research in the area of cell, human, plant and animal physiology.

Meet the Series Editor

Prof. Dr. Thomas Brzozowski works as a professor of Human Physiology and is currently a Chairman at the Department of Physiology and is V-Dean of the Medical Faculty at Jagiellonian University Medical College, Cracow, Poland. His primary area of interest is physiology and pathophysiology of the gastrointestinal (GI) tract, with a major focus on the mechanism of GI mucosal defense, protection, and ulcer healing. He was a postdoctoral NIH fellow at the University of California and the Gastroenterology VA Medical Center, Irvine, Long Beach, CA, USA, and at the Gastroenterology Clinics Erlangen-Nuremberg and Munster in Germany. He has published 290 original articles in some of the most prestigious scientific journals and seven book chapters on the pathophysiology of the GI tract, gastroprotection, ulcer healing, drug therapy of peptic ulcers, hormonal regulation of the gut, and inflammatory bowel disease.

Meet the Volume Editor

Dr. Hassan Massoud Heshmati is an endocrinologist with 46 years of experience in clinical research in academia (university-affiliated hospitals, Paris, France; Mayo Foundation, Rochester, MN, USA) and pharmaceutical companies (Sanofi, Malvern, PA, USA; Essentialis, Carlsbad, CA, USA; Gelesis, Boston, MA, USA). His research activity focuses on pituitary tumors, hyperthyroidism, thyroid cancers, osteoporosis, diabetes, and obesity. He has extensive knowledge in the development of anti-obesity products. Dr. Heshmati is the author of 299 abstracts, chapters, and articles related to endocrinology and metabolism. He is currently a consultant at Endocrinology Metabolism Consulting, LLC, Anthem, AZ, USA.

Contents

Preface

In biology, senescence is a process by which cell ages and permanently stops dividing. Senescence is a natural universal phenomenon affecting all living organisms (e.g., humans, animals, and plants). The underlying mechanisms of senescence at the cellular level are not fully understood. Senescence is a multifactorial process that can be induced by several stimuli including cellular stress, DNA damage, telomere shortening, and oncogene activation. Senescence impacts the lifespan of living organisms. The lifespan ranges from a few hours (with mayfly) to potential eternity (with jellyfish and hydra). The maximum theoretical lifespan in humans is around 120 years. Senescence plays a role in the development of several age-related chronic diseases in humans (e.g., ischemic heart disease, osteoporosis, and cancer). Senescence is emerging as a therapeutic target for several diseases. Clinical trials using anti-senescent therapies are in progress. Results from early pilot studies suggest that senolytic drugs can decrease senescent cells, reduce inflammation, and alleviate frailty. This book provides the reader with a comprehensive overview of the current knowledge on the mechanisms and the management of senescence in humans together with comparative data on animals and plants.

The book contains six chapters by authors from Russia, Serbia, and the United States. I would like to thank all of them. I would also like to thank the great assistance of Ms. Dolores Kuzelj at IntechOpen who supervised this book project.

Hassan M. Heshmati, M.D.
Endocrinology Metabolism Consulting, LLC,
Anthem, AZ, USA

Section 1

Universality of Senescence

Chapter 1

Comparative Senescence and Lifespan

Hassan M. Heshmati

Abstract

The word senescence is derived from the Latin word "senex" (meaning old). In biology, senescence is a process by which a cell ages and permanently stops dividing. Senescence is a natural universal phenomenon affecting all living organisms (e.g., humans, animals, and plants). It is the process of growing old (aging). The underlying mechanisms of senescence and aging at the cellular level are not fully understood. Senescence is a multifactorial process that can be induced by several stimuli including cellular stress, DNA damage, telomere shortening, and oncogene activation. The most popular theory to explain aging is the free radical theory. Senescence plays a role in the development of several age-related chronic diseases in humans (e.g., ischemic heart disease, osteoporosis, and cancer). Lifespan is a biological characteristic of every species. The lifespan of living organisms ranges from few hours (with mayfly) to potential eternity (with jellyfish and hydra). The maximum theoretical lifespan in humans is around 120 years. The lifespan in humans is influenced by multiple factors including genetic, epigenetic, lifestyle, environmental, metabolic, and endocrine factors. There are several ways to potentially extend the lifespan of humans and eventually surpass the maximum theoretical lifespan of 120 years. The tools that can be proposed include lifestyle, reduction of several life-threatening diseases and disabilities, hormonal replacement, antioxidants, autophagy inducers, senolytic drugs, stem cell therapy, and gene therapy.

Keywords: senescence, aging, lifespan, humans, animals, plants

1. Introduction

Senescence is a process by which a cell ages and permanently stops dividing [1]. It is a natural universal phenomenon affecting all living organisms (e.g., humans, animals, and plants). Senescence in cells occurs with the process of growing old (aging). It is a multifactorial process that can be induced by several stimuli including cellular stress, DNA damage, telomere shortening, and oncogene activation.

Lifespan is a biological characteristic of every species. The lifespan of living organisms ranges from few hours (with mayfly) to potential eternity (with jellyfish and hydra). The maximum theoretical lifespan in humans is around 120 years. The lifespan in humans is influenced by multiple factors including genetic, epigenetic, lifestyle, environmental, metabolic, and endocrine factors [2–26].

Lifespan can potentially be extended by mutations or by a variety of interventions in humans, animals, and plants. In humans, interventions such as lifestyle, reduction of several life-threatening diseases and disabilities, hormonal replacement, antioxidants, autophagy inducers, senolytic drugs, stem cell therapy, and gene therapy can potentially help to extend lifespan and eventually surpass 120 years [6–8, 12–15, 17, 20–22, 24–41].

This chapter presents a comparative overview of senescence and lifespan in humans, animals, and plants.

2. History of life on planet Earth

According to the core accretion theory, planet Earth formed around 4.54 billion years ago (approximately one-third the age of the universe) by accretion from the solar nebula [42]. The initial Earth was a dry planet, without atmosphere and ocean components [43].

The origin of life question is one of the most challenging questions in science. Both ribonucleic acid and peptides played key roles in the emergence of life on Earth [44]. Although the beginning of the presence of life on Earth cannot be determined with accuracy, there is evidence that primitive life with bacteria-like organisms was present around 3.50 billion years ago (**Figure 1**).

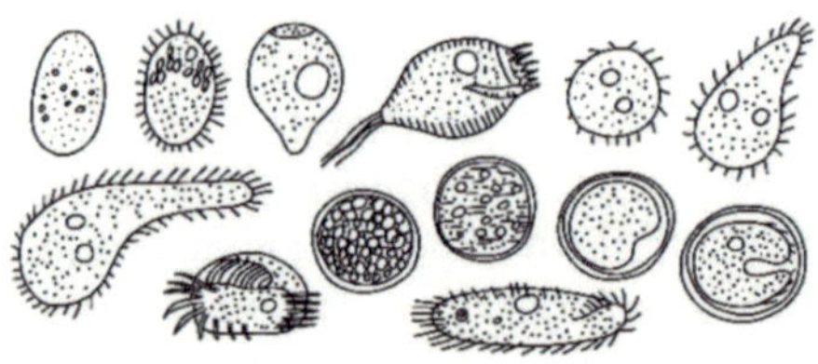

Figure 1.
Primitive life was likely present on Earth around 3.50 billion years ago. Copyright symkin (Николай Бибик)/ Depositphotos Inc.

The evolution of species is a complex phenomenon that is supported by several theories. According to Darwin, the natural selection enabled simple life to evolve into complex life over a long period of time (**Figure 2**) [45].

Figure 2.
Evolution of humans. Copyright YAYImages (Alex Golke)/Depositphotos Inc.

3. Diversity of life on planet Earth

Diversity of life refers to the variety of existing living organisms in relation to their species, races, ethnicities, and habitats.

3.1 Species, races, and ethnicities

There is a variety of species (several million), races, and ethnicities on planet Earth. However, most species, including those living in the ocean, remain to be described [46]. Some species are extremely abundant, while others are moderately common or rare.

3.1.1 Humans

The historical definition of human races has evolved over more than two centuries. The initial definition of Black, Brown, Red, White, and Yellow races was at the origin of prejudice and racism. These shameful attitudes are still present in our societies.

To encourage diversity and promote uniformity and comparability of data on race and ethnicity, the National Institutes of Health in the USA has defined the following racial and ethnic categories: American Indian or Alaska Native, Asian, Black or African American, Hispanic or Latino, and White. To be comprehensive with this classification, we should also include different combinations of the above races and ethnicities (**Figure 3**).

Figure 3.
Diversity of humans. Copyright macrovector (Alejandro Sánchez Blanes)/Depositphotos Inc.

3.1.2 Animals

It is estimated that there are over 7 million animal species, but most have not been described (**Figure 4**). With approximately 1,000,000 described species, insects represent about 55% of all known species.

The basic animal classes include invertebrates, fish, amphibians, reptiles, birds, and mammals. Within each animal species, there are several breeds. For dogs, for example, more than 300 breeds have been described.

Figure 4.
Diversity of animal species. Copyright Jim_Filim (Iakov Filimonov)/Depositphotos Inc.

3.1.3 Plants

There are close to 400,000 species of plants. The large majority of plants are vascular and flowering plants (**Figure 5**).

Figure 5.
Diversity of plant species. Copyright interactimages (Matthew Cole)/Depositphotos Inc.

3.2 Habitats

There are two main types of habitats for the living organisms: terrestrial (e.g., regular land, grassland, forest, desert, mountain, and polar region) and aquatic (freshwater, marine, and coastal region). Based on their habitat, living organisms can be exposed to a variety of air, sunlight, temperature, humidity, and noise (**Figure 6**).

Figure 6.
Organisms living in the water. Copyright Vlad61 (Владимир Голубев)/Depositphotos Inc.

4. Senescence and aging

The word senescence is derived from the Latin word "senex" (meaning old). In biology, senescence is a process by which a cell ages and permanently stops dividing. In addition to exiting the cell cycle, senescent cells undergo other phenotypic alterations including metabolic reprogramming, chromatin rearrangement, and autophagy modulation [1]. Senescence is a natural universal phenomenon affecting all living organisms (e.g., humans, animals, and plants). However, there is a wide range of diversity in the pattern of senescence across species. Senescence is the process of growing old (aging) with a progressive deterioration of the cell and organ functioning. Senescence is associated with a decrease in fertility and/or an increase in mortality.

4.1 Mechanisms

The underlying mechanisms of senescence and aging at the cellular level are not fully understood. Senescence is a multifactorial process that can be induced by several stimuli including cellular stress, DNA damage, telomere shortening, and oncogene activation. The most popular theory to explain aging is the free radical theory [47]. According to this theory, continuous, unrepaired oxidative damage of macromolecules constitutes the molecular basis of aging.

4.1.1 Humans

Human senescence and aging result from accumulation over time of genetic, molecular, and cellular damages.

Aging is associated with a gradual, time-dependent, and heterogeneous decline of physiological functions ultimately leading to death (**Figure** 7). The human body goes through multiple changes including endothelial pro-atherosclerotic changes, an overall decrease in the size of organs, ovarian atrophy, osteopenia, sarcopenia, skin atrophy, and adipose tissue enlargement (non-exhaustive list) [7, 10, 12, 26]. Some of these changes play a role in the development of several age-related chronic diseases (e.g., ischemic heart disease, osteoporosis, and cancer) responsible for increased mortality. Senescence is emerging as a therapeutic target for several diseases.

Figure 7.
Evolution of humans with aging. Copyright cherstva (Olena Cherstva)/Depositphotos Inc.

4.1.2 Animals

Senescence and aging are also present in animals and influenced by multiple factors including genetic and environmental factors [48]. The rate and intensity of aging vary considerably among animals. Interestingly, some animals do not exhibit clear evidence of biological aging (negligible senescence) and have superior resistance to age-related diseases. These animals include ocean quahog clam, Greenland shark, Aldabra giant tortoise, rougheye rockfish, freshwater pearl mussel, and naked mole-rat (non-exhaustive list). At least two animals, jellyfish (*Turritopsis dohrnii*) and hydra, are considered potentially immortal [49].

4.1.3 Plants

Senescence and aging in plants are associated with a complex deterioration of cellular metabolism that includes loss of chlorophyll, carotenoids, and proteins, and an increase in lipid peroxidation and membrane permeability, leading to a decline in photosynthesis. Multiple factors including phytohormones, sunlight, temperature, and water play a role in plant senescence [50].

5. Lifespan

Lifespan is a biological characteristic of every species. It is determined by a complex interaction between genetic and environmental factors. The lifespan of living organisms ranges from few hours (animals) to few thousand years (animals and plants) or to eternity (animals). Phenotypic plasticity can affect the long lifespan of both animals and plants [48].

5.1 Lifespan by species

5.1.1 Humans

The lifespan in humans is influenced by multiple factors including genetic, epigenetic, lifestyle, environmental, metabolic, and endocrine factors [2–26].

The maximum theoretical lifespan in humans is around 120 years. Very few individuals reach this theoretical age since several events can impact lifespan (e.g., diseases, suicide, accident, and war).

Centenarians are subjects living 100 years or older. They represent a model of successful aging [9, 31, 51, 52]. Semi-supercentenarians are those who reach an age of 105–109 years (**Figure 8**). A very small fraction of centenarians (up to 0.5%) will live 110 years or older (supercentenarians) [5, 52]. The oldest supercentenarian with well-documented age was Jeanne Louise Calment (1875–1997) from Arles (France) who lived 122 years. The second oldest supercentenarian was Sarah DeRemer Knauss (1880–1999) from Hollywood, Pennsylvania (USA) who lived 119 years.

According to United Nations estimates, in 2020, the number of centenarians in the world was approximately 573,000 (mainly from the USA). This number could reach approximately 3,676,000 by 2050 (mainly from China).

Figure 8.
A semi-supercentenarian Chinese woman (105 years old) and her great granddaughter. Copyright ChinaImages (IMAGINECHINA LIMITED)/Depositphotos Inc.

5.1.2 Animals

The lifespan in animals is between few hours to potential eternity. The shortest lifespan is seen with mayfly (1 day). Aside from jellyfish (*Turritopsis dohrnii*) (**Figure 9**) and hydra which are considered potentially immortal animals, the longest lifespan has been observed with glass sponge (around 10,000 years) (**Table 1**) [49].

Figure 9.
Jellyfish (Turritopsis dohrnii) is considered a potentially immortal animal. Copyright Winston (Sergey Belov)/ Depositphotos Inc.

Animal	**Lifespan**
Jellyfish	Potentially immortal
Hydra	Potentially immortal
Glass sponge	10,000 years
Coral	4,000 years
Clam	500 years
Shark	270 years
Giant tortoise	250 years
Rougheye rockfish	200 years
Parrot	100 years
Crocodile	70 years
Elephant	50 years
Monkey	45 years
Camel	40 years
Horse	30 years
Cow	20 years
Snake	20 years
Cat	18 years
Lion	15 years
Tiger	15 years
Dog	13 years

Animal	Lifespan
Rabbit	12 years
Chicken	10 years
Pigeon	5 years
Ant	4 years
Rat	2 years
Mosquito	6 weeks
Fly	1 month
Mayfly	1 day

Table 1.
Lifespan (average or record) in animals ranked by descending order of duration (non-exhaustive list).

5.1.3 Plants

The lifespan in plants ranges from few weeks (in annuals) to few thousand years (in trees). Bristlecone pine (*Pinus longaeva*), a tree found in the higher mountains of California (USA), is among the plants with the longest lifespan (around 5,000 years) (**Figure 10**).

Figure 10.
Bristlecone pine (Pinus longaeva) can live around 5,000 years. Copyright mkopka (Melissa Kopka)/ Depositphotos Inc.

5.2 Evolution of lifespan

The important increase in human lifespan over the past 100 years is one of the greatest achievements of humanity [53]. The evolution of human lifespan over time is reported in **Table 2**.

Period	Lifespan
Pre-historic	30 years
Early 16th century (developed countries)	40 years
Early 20th century (developed countries)	50 years
Early 21st century (worldwide)	73 years

Table 2.
Evolution of average lifespan in humans throughout the history of mankind.

5.3 Interventions to extend lifespan

Although lifespan is a biological characteristic of every species, it can be modified by mutations or by a variety of interventions in humans, animals, and plants.

Extending lifespan while keeping health and vitality has always been a dream for mankind. The "successful aging" is aging without any disabilities and severe diseases (**Figure 11**) [27, 53].

Figure 11.
Healthy old couple. Copyright nicoletaionescu (Nicoleta Ionescu)/Depositphotos Inc.

There are several ways to potentially extend the lifespan of humans and eventually surpass the maximum theoretical lifespan of 120 years. The tools that can be proposed include lifestyle, reduction of several life-threatening diseases and disabilities, hormonal replacement, antioxidants, autophagy inducers, senolytic drugs, stem cell therapy, and gene therapy (**Table 3**) (**Figure 12**) [6–8, 12–15, 17, 20–22, 24–41].

Clinical trials using anti-senescent therapies are in progress. Results from early pilot studies suggest that senolytic drugs can decrease senescent cells, reduce inflammation, and alleviate frailty. Stem cell therapy represents a new emerging era in medicine that has the potential to delay the aging process and, therefore, extend lifespan, by better treating life-threatening diseases that impact lifespan. Genetic interventions, although promising, may be difficult to implement in humans without the knowledge of all the potential health consequences during entire life.

Tool	Example
Lifestyle	Diet (rich in vegetables, fruits, and fiber, poor in saturated fat and red meat), Caloric restriction, Exercise, Clean and stress-free environment
Reduction of life-threatening diseases	Treatment of Ischemic heart disease, Chronic obstructive pulmonary disease, Cancer, Overweight/Obesity, Type 2 diabetes, Osteoporosis, etc.
Hormonal replacement	Estrogen, Dehydroepiandrosterone
Antioxidants	Resveratrol, Curcumin, etc.
Autophagy inducers	Caloric restriction, Exercise, Vitamin D, Resveratrol, Metformin, Rapamycin, etc.
Senolytic drugs	Dasatinib, Quercetin, Fisetin, Navitoclax, etc.
Stem cell therapy	Stem cells
Gene therapy	Gene editing, Viral or Non-viral vectors

Table 3.
Tools proposed for the potential extension of the lifespan in humans (non-exhaustive list).

Figure 12.
Regular exercise can be very beneficial for the extension of lifespan. Copyright kornetka (Svetlana Ivanova)/ Depositphotos Inc.

6. Conclusions

Senescence is a process by which a cell ages and permanently stops dividing. It is a natural universal phenomenon affecting all living organisms (e.g., humans, animals, and plants). Senescence is the process of growing old (aging). It can be induced by several stimuli including cellular stress, DNA damage, telomere shortening, and oncogene activation. The most popular theory to explain aging is the free radical theory.

Lifespan is a biological characteristic of every species. The lifespan of living organisms ranges from few hours (with mayfly) to potential eternity (with jellyfish and hydra). The maximum theoretical lifespan in humans is around 120 years. The lifespan in humans is influenced by multiple factors including genetic, epigenetic, lifestyle, environmental, metabolic, and endocrine factors. The lifespan in animals is between few hours to potential eternity. The lifespan in plants ranges from few weeks to few thousand years.

There are tools that can potentially extend the lifespan of humans and eventually surpass 120 years. They include lifestyle, reduction of several life-threatening diseases and disabilities, hormonal replacement, antioxidants, autophagy inducers, senolytic drugs, stem cell therapy, and gene therapy.

Conflict of interest

The author declares no conflict of interest.

Author details

Hassan M. Heshmati
Endocrinology Metabolism Consulting, LLC, Anthem, AZ, USA

*Address all correspondence to: hassanheshmati@yahoo.com

References

[1] Herranz N, Gil J. Mechanisms and functions of cellular senescence. The Journal of Clinical Investigation. 2018;**128**:1238-1246. DOI: 10.1172/JCI95148

[2] Sebastiani P, Perls TT. The genetics of extreme longevity: Lessons from the New England Centenarian Study. Frontiers in Genetics. 2012;**3**:277. DOI: 10.3389/fgene.2012.00277

[3] Budovsky A, Craig T, Wang J, et al. LongevityMap: A database of human genetic variants associated with longevity. Trends in Genetics. 2013;**29**:559-560. DOI: 10.1016/j.tig.2013.08.003

[4] Shadyab AH, LaCroix AZ. Genetic factors associated with longevity: A review of recent findings. Ageing Research Reviews. 2015;**19**:1-7. DOI: 10.1016/j.arr.2014.10.005

[5] Horvath S, Pirazzini C, Bacalini MG, et al. Decreased epigenetic age of PBMCs from Italian semi-supercentenarians and their offspring. Aging. 2015;7:1159-1170. DOI: 10.18632/aging.100861

[6] Govindaraju D, Atzmon G, Barzilai N. Genetics, lifestyle and longevity: Lessons from centenarians. Applied & Translational Genomics. 2015;**4**:23-32. DOI: 10.1016/j.atg.2015.01.001

[7] Sgarbieri VC, Pacheco MTB. Healthy human aging: Intrinsic and environmental factors. Brazilian Journal of Food Technology. 2017;**20**:e2017007. DOI: 10.1590/1981-6723.00717

[8] Puca AA, Spinelli C, Accardi G, Villa F, Caruso C. Centenarians as a model to discover genetic and epigenetic signatures of healthy ageing. Mechanisms of Ageing and Development. 2018;**174**:95-102. DOI: 10.1016/j.mad.2017.10.004

[9] Borras C, Ingles M, Mas-Bargues C, et al. Centenarians: An excellent example of resilience for successful ageing. Mechanisms of Ageing and Development. 2020;**186**:111199. DOI: 10.1016/j.mad.2019.111199

[10] Bektas A, Schurman SH, Sen R, Ferrucci L. Aging, inflammation and the environment. Experimental Gerontology. 2018;**105**:10-18. DOI: 10.1016/j.exger.2017.12.015

[11] Barzilai N, Gabriely I. Genetic studies reveal the role of the endocrine and metabolic systems in aging. The Journal of Clinical Endocrinology & Metabolism. 2010;**95**:4493-4500. DOI: 10.1210/jc.2010-0859

[12] Barzilai N, Huffman DM, Muzumdar RH, Bartke A. The critical role of metabolic pathways in aging. Diabetes. 2012;**61**:1315-1322. DOI: 10.2337/db11-1300

[13] López-Otín C, Galluzzi L, Freije JMP, Madeo F, Kroemer G. Metabolic control of longevity. Cell. 2016;**166**:802-821. DOI: 10.1016/j.cell.2016.07.031

[14] Barbosa MC, Grosso RA, Fader CM. Hallmarks of aging: An autophagic perspective. Frontiers in Endocrinology. 2019;**9**:790. DOI: 10.3389/fendo.2018.00790

[15] Allard JB, Duan C. Comparative endocrinology of aging and longevity regulation. Frontiers in Endocrinology. 2011;**2**:75. DOI: 10.3389/fendo.2011.00075

[16] Chen TT, Maevsky EI, Uchitel ML. Maintenance of homeostasis in the aging

hypothalamus: The central and peripheral roles of succinate. Frontiers in Endocrinology. 2015;**6**:7. DOI: 10.3389/fendo.2015.00007

[17] Diamanti-Kandarakis E, Dattilo M, Macut D, et al. Aging and anti-aging: A combo-endocrinology overview. European Journal of Endocrinology. 2017;**176**:R283-R308. DOI: 10.1530/EJE-16-1061

[18] Colon G, Saccon T, Schneider A, et al. The enigmatic role of growth hormone in age-related diseases, cognition, and longevity. GeroScience. 2019;**41**:759-774. DOI: 10.1007/s11357-019-00096-w

[19] Bartke A. Growth hormone and aging. Reviews in Endocrine and Metabolic Disorders. 2021;**22**:71-80. DOI: 10.1007/s11154-020-09593-2

[20] Bowers J, Terrien J, Clerget-Froidevaux MS, et al. Thyroid hormone signaling and homeostasis during aging. Endocrine Reviews. 2013;**34**:556-589. DOI: 10.1210/er.2012-1056

[21] Longo M, Bellastella G, Maiorino MI, Meier JJ, Esposito K, Giugliano D. Diabetes and aging: From treatment goals to pharmacologic therapy. Frontiers in Endocrinology. 2019;**10**:45. DOI: 10.3389/fendo.2019.00045

[22] Salpeter SR, Cheng J, Thabane L, Buckley NS, Salpeter EE. Bayesian meta-analysis of hormone therapy and mortality in younger postmenopausal women. The American Journal of Medicine. 2009;**122**:1016-1022. DOI: 10.1016/j.amjmed.2009.05.021

[23] Araujo AB, Dixon JM, Suarez EA, Hassan Murad M, Guey LT, Wittert GA. Endogenous testosterone and mortality in men: A systematic review and meta-analysis. The Journal of Clinical Endocrinology & Metabolism. 2011;**96**:3007-3019. DOI: 10.1210/jc.2011-1137

[24] Lorenzini A. How much should we weigh for a long and healthy life span? The need to reconcile caloric restriction versus longevity with body mass index versus mortality data. Frontiers in Endocrinology. 2014;**5**:121. DOI: 10.3389/fendo.2014.00121

[25] The Global BMI Mortality Collaboration. Body-mass index and all-cause mortality: Individual-participant-data meta-analysis of 239 prospective studies in four continents. Lancet. 2016;**388**:776-786. DOI: 10.1016/S0140-6736(16)30175-1

[26] Mancuso P, Bouchard B. The impact of aging on adipose function and adipokine synthesis. Frontiers in Endocrinology. 2019;**10**:137. DOI: 10.3389/fendo.2019.00137

[27] Beard JR, Officer A, Araujo de Carvalho I, et al. The world report on ageing and health: A policy framework for healthy ageing. Lancet. 2016;**387**:2145-2154. DOI: 10.1016/S0140-6736(15)00516-4

[28] Barzilai N, Cuervo AM, Austad S. Aging as a biological target for prevention and therapy. Journal of the American Medical Association. 2018;**320**:1321-1322. DOI: 10.1001/jama.2018.9562

[29] Shetty AK, Kodali M, Upadhya R, Madhu LN. Emerging anti-aging strategies – Scientific basis and efficacy. Aging and Disease. 2018;**9**:1165-1184. DOI: 10.14336/AD.2018.1026

[30] Tchkonia T, Kirkland JL. Aging, cell senescence, and chronic disease. Emerging therapeutic strategies.

Journal of the American Medical Association. 2018;**320**:1319-1320. DOI: 10.1001/jama.2018.12440

[31] Heshmati HM. The centenarians: An emerging population. In: Amornyotin S, editor. Update in Geriatrics. London: IntechOpen; 2021. pp. 3-23. DOI: 10.5772/intechopen.96327

[32] Abbasi J. Can a diet that mimics fasting turn back the clock? Journal of the American Medical Association. 2017;**318**:227-229. DOI: 10.1001/jama.2017.6648

[33] Pereira da Silva A, Valente A, Chaves C, et al. Characterization of Portuguese centenarian eating habits, nutritional biomarkers, and cardiovascular risk: A case control study. Oxidative Medicine and Cellular Longevity. 2018;**2018**:5296168. DOI: 10.1155/2018/5296168

[34] Castillo-Garzón MJ, Ruiz JR, Ortega FB, Gutiérrez ÁG. Anti-aging therapy through fitness enhancement. Clinical Interventions in Aging. 2006;**1**:213-220. DOI: 10.2147/ciia.2006.1.3.213

[35] Weiss EP, Villareal DT, Fontana L, Han DH, Holloszy JO. Dehydroepiandrosterone (DHEA) replacement decreases insulin resistance and lowers inflammatory cytokines in aging humans. Aging. 2011;**3**:533-542. DOI: 10.18632/aging.100327

[36] Levine B, Packer M, Codogno P. Development of autophagy inducers in clinical medicine. The Journal of Clinical Investigation. 2015;**125**:14-24. DOI: 10.1172/JCI73938

[37] Myrianthopoulos V. The emerging field of senotherapeutic drugs. Future Medicinal Chemistry. 2018;**10**:2369-2372. DOI: 10.4155/fmc-2018-0234

[38] Kirkland JL, Tchkonia T. Senolytic drugs: From discovery to translation. Journal of Internal Medicine. 2020;**288**:518-536. DOI: 1111/joim.13141

[39] Ullah M, Sun Z. Stem cells and anti-aging genes: Double-edged sword-do the same job of life extension. Stem Cell Research & Therapy. 2018;**9**:3. DOI: 10.1186/s13287-017-0746-4

[40] Schulman IH, Balkan W, Hare JM. Mesenchymal stem cell therapy for aging frailty. Frontiers in Nutrition. 2018;**5**:108. DOI: 10.3389/fnut.2018.00108

[41] Vaiserman A, De Falco E, Koliada A, Maslova O, Balistreri CR. Anti-ageing gene therapy: Not so far away? Ageing Research Reviews. 2019;**56**:100977. DOI: 10.1016/j.arr.2019.100977

[42] Co'Neill HS. The origin of the moon and the early history of the earth – A chemical model. Part 2: The earth. Geochimica et Cosmochimica Acta. 1991;**55**:1159-1172. DOI: 10.1016/0016-7037(91)90169-6

[43] Maruyama S, Ebisuzaki T. Origin of the Earth: A proposal of new model called ABEL. Geoscience Frontiers. 2017;**8**:253-274. DOI: 10.1016/j.gsf.2016.10.005

[44] Saad NY. A ribonucleopeptide world at the origin of life. Journal of Systematics and Evolution. 2018;**56**:1-13. DOI: 10.1111/jse.12287

[45] Avise JC, Ayala FJ. In the light of evolution III: Two centuries of Darwin. PNAS. 2009;**106**(Suppl 1):9933-9938. DOI: 10.1073/pnas.0903381106

[46] Mora C, Tittensor DP, Adl S, Simpson AGB, Worm B. How many species are there on Earth and in the ocean? PLOS Biology. 2011;**9**:e1001127. DOI: 10.1371/journal.pbio.1001127

[47] Harman D. Aging: Origin and evolution of the free radical theory of aging: A brief personal history, 1954-2009. Biogerontology. 2009;**10**:773-781. DOI: 10.1007/s10522-009-9234-2

[48] Borges RM. Phenotypic plasticity and longevity in plants and animals: Cause and effect? Journal of Biosciences. 2009;**34**:605-611. DOI: 10.1007/s12038-009-0078-3

[49] Petralia RS, Mattson MP, Yao PJ. Aging and longevity in the simplest animals and the quest for immortality. Ageing Research Reviews. 2014;**16**: 66-82. DOI: 10.1016/j.arr.2014.05.003

[50] Jyothsna P, Murthy SDS. A review on effect of senescence in plants and role of phytohormones in delaying senescence. International Journal of Plant, Animal and Environmental Sciences. 2016;**6**:152-161

[51] Salvioli S, Capri M, Bucci L, et al. Why do centenarians escape or postpone cancer? The role of IGF-1, inflammation and p53. Cancer Immunology, Immunotherapy. 2009;**58**:1909-1917. DOI: 10.1007/s00262-008-0639-6

[52] Andersen SL, Sebastiani P, Dworkis DA, Feldman L, Perls TT. Health span approximates life span among many supercentenarians: Compression of morbidity at the approximate limit of life span. The Journals of Gerontology, Series A: Biological Sciences and Medical Sciences. 2012;**67A**:395-405. DOI: 10.1093/gerona/glr223

[53] Olshansky SJ. From lifespan to healthspan. Journal of the American Medical Association. 2018;**320**:1323-1324. DOI: 10.1001/jama.2018.12621

Section 2

Senescence in Cells and Organs

Chapter 2

Identification of RNA Species That Bind to the hnRNP A1 in Normal and Senescent Human Fibroblasts

Heriberto Moran, Shanaz A. Ghandhi, Naoko Shimada and Karen Hubbard

Abstract

hnRNP A1 is a member of the hnRNPs (heterogeneous nuclear ribonucleoproteins) family of proteins that play a central role in regulating genes responsible for cell proliferation, DNA repair, apoptosis, and telomere biogenesis. Previous studies have shown that hnRNPA1 had reduced protein levels and increased cytoplasmic accumulation in senescent human diploid fibroblasts. The consequence of reduced protein expression and altered cellular localization may account for the alterations in gene expression observed during senescence. There is limited information for gene targets of hnRNP A1 as well as its *in vivo* function. In these studies, we performed RNA co-immunoprecipitation experiments using hnRNP A1 as the target protein to identify potential mRNA species in ribonucleoprotein (RNP) complexes. Using this approach, we identified the human double minute 2 (*HDM2)* mRNA as a binding target for hnRNP A1 in young and senescent human diploid fibroblasts cells. It was also observed that alterations of hnRNP A1 expression modulate *HDM2* mRNA levels in young IMR-90 cells. We also demonstrated that the levels of *HDM2* mRNA increased with the downregulation of hnRNP A1 and decrease with the overexpression of hnRNP A1. Although we did not observe a significant decrease in HDM2 protein level, a concomitant increase in p53 protein level was detected with the overexpression of hnRNP A1. Our studies also show that hnRNP A1 directly interacts with HDM2 mRNA at a region corresponding to its 3′ UTR (untranslated region of a gene). The results from this study demonstrate that hnRNP A1 has a novel role in participating in the regulation of HDM2 gene expression.

Keywords: senescence, fibroblasts, hnRNP A1, *hdm2*, RNP complexes

1. Introduction

Cellular senescence is best described as an inevitable irreversible proliferation arrest the phase of primary human fibroblasts in culture [1, 2]. The senescent phenotype is characterized by distinctive changes in morphology to become enlarged, flattened, and granular [2]. Multiple factors that activate senescence include various types of stress-related stimuli such as aberrant oncogenic signaling, oxidative stress, and DNA damage [2]. Moreover, the onset of senescence can be regulated by events such as epigenetic regulation, chromosome dynamics, protein degradation, mitochondrial mechanisms, and metabolic pathways. The molecular pathways of senescence differ considerably among cell types as well as different species [3].

Alternative splicing of pre-mRNAs is a process in which varied mRNA transcripts are generated to provide a major source of protein diversity in higher eukaryotes. Pre-mRNA splicing is a nuclear process that can be constitutive or alternative [4, 5]. Constitutive splicing involves the removal of introns and the joining of adjacent exons in the order of their arrangement. One of the core proteins involved in splicing is hnRNPA1 [5, 6]. Consequently, a single protein may be produced from a single pre-mRNA in constitutive splicing [7]. In contrast, in alternative splicing, the variable use of splice sites permits two or more mature mRNAs to be generated from the same pre-mRNA. Among the nuclear complexes primarily responsible for alternative splicing are heterogeneous nuclear ribonucleoproteins, small nuclear ribonucleoproteins snRNPs, and SR proteins [8, 9]. Splicing factors that play a crucial role through concentration changes or alterations of their expression patterns have significant impacts on mRNA alternative splicing [9, 10].

We have previously found that hnRNP A1 is significantly downregulated in cellular senescence [10] and can regulate the levels of the alternatively spliced INK^{4a} locus that generates the mRNA isoforms, $p16^{INK4a}$ and $p14^{ARF}$ both of which are growth suppressors that are important in senescence [10, 11]. Increased expression levels of hnRNP A1 *via* over-expression can shift the expression pattern toward the $p14^{ARF}$ mRNA isoform [10].

We initiated this study to identify novel targets of hnRNP A1 and to further explore the role of hnRNP A1 in the modulation of gene expression during cellular senescence. The experimental approach used in this study was to identify the *in vivo* RNA targets bound in hnRNP A1 RNP complexes isolated from human fibroblasts. RNP complexes were isolated by a brief co-immunoprecipitation step with the 4B10 hnRNP A1-specific monoclonal antibody [12]. RNA species in these complexes were then subjected to reverse transcription followed by amplification. The products were then cloned and sequenced. Our findings suggest that hnRNP A1 is involved in the regulation of *HDM2* gene expression. However, the involvement of hnRNP A1 in the regulation of *HDM2* gene expression remains to be elucidated.

2. Results

2.1 hnRNP A1-messenger ribonucleoprotein (mRNP) complexes

We sought to identify putative mRNA substrates for hnRNP A1 by identifying mRNA sequences that directly bind to the hnRNP A1 protein. We employed a modification of a procedure that had been used for the characterization of RNP complexes by Mili et al. [12]. We isolated hnRNP A1 protein complexes bound to their RNA targets from total young and senescent fibroblast cell lysates. To demonstrate that the complexes represented the majority of hnRNP A1-mRNP complexes found in cellular pools, we measured the mRNA levels of hnRNP A1 and actin in the isolated complexes by RT-PCR. Actin has been previously reported to be in hnRNP A1 RNP complexes; thus, we used actin as a positive control. The results in **Figure 1A**, show that actin and hnRNP A1 protein were present in lysates isolated from young and senescent IMR-90 fibroblasts. We assessed the protein level of hnRNP A1 in 4B10 RNP complexes

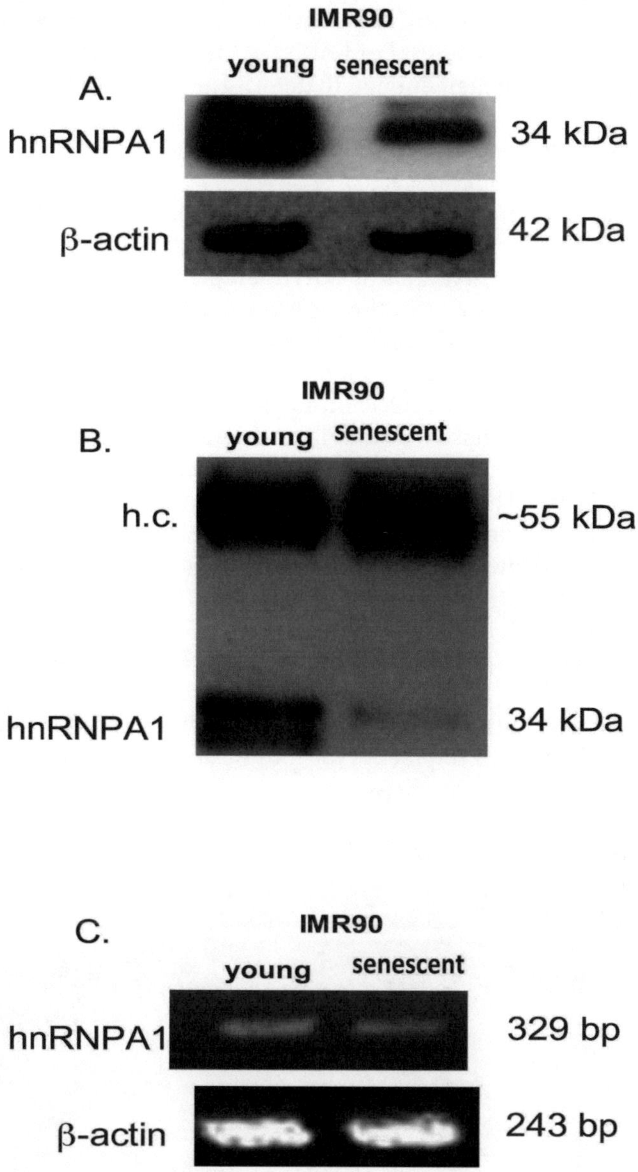

Figure 1.
A. Expression of hnRNPA1 in young and old IMR90 fibroblasts. The endogenous level of hnRNP A1 protein is significantly lower in old IMR90 fibroblasts (major band at ~34 kDa is hnRNPA1). Equivalent amounts of protein (20 μg) were loaded. B. Post-immunoprecipitation levels of hnRNPA1 were also low in old cells, see the band at ~34 kDa. C. RT-PCR on cDNA transcribed from RNA isolated complexes, using hnRNPA1 and β-actin primer sets. In both young and old lysates, there appear to be comparable amounts of hnRNPA1 mRNA species in the isolated hnRNPA1-containing complexes.

isolated from young and senescent protein lysates. 4B10 RNP complexes reflect the results in panel A (**Figure 1B**) as hnRNP A1 was lower in senescent IMR-90 protein lysates following immunoprecipitation when compared with young cell lysates. These observations indicate that hnRNP A1 was present in the isolated RNP complexes. One known mRNA target of hnRNP A1 is its own RNA; therefore, we measured the level of hnRNP A1 mRNA by RT-PCR (**Figure 1C**). We found that the 4B10 monoclonal antibody immune-precipitated hnRNP A1 mRNA in RNP complexes from both young and senescent lysates. While the level of hnRNP A1 mRNA was lower in senescent RNP complexes, it was sufficient enough to determine putative mRNA targets.

2.2 Analysis of the hnRNPA1-mRNA complex

RNA that was isolated from the hnRNP A1-complexes was reverse transcribed and then amplified by PCR using random decamers to amplify all cDNA sequences. We used two different concentrations of cDNA template for PCR as it was not always possible to visualize PCR products in the senescent samples because the cDNA abundance is typically lower in these cells. There was a correlative increase in PCR products as shown in **Figure 2** with an increase in cDNA template. PCR products were then immediately ligated into a pCR2.1cloning vector.

There may also be a differential availability of hnRNP A1 protein in old cells as hnRNPA1 is modified by post-translational events, such as phosphorylation and methylation [13, 14]. There is an additional possibility that rearrangements of individual components in hnRNPA1-mRNA ribonucleoparticles may change during senescence, which could alter specific and non-specific mRNA sequences bound in the complex.

2.3 Identification of RNA species in hnRNP A1-mRNP complexes

We then determined the identity of the cloned inserts by sequencing. We found that there were partial mRNA sequences for four human genes bound in young and

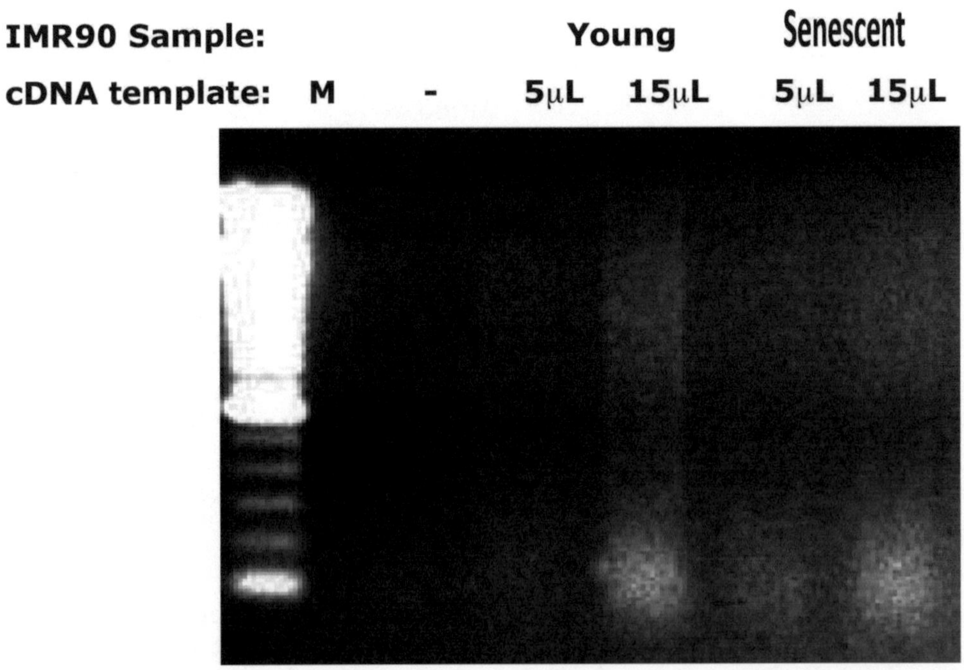

Figure 2.
Analysis of hnRNP A1 RNP complexes. A. RNA isolated from hnRNPA1 complexes was reverse transcribed to cDNA and amplified using random decamers as primers for the reaction. No amplification was observed in the absence of cDNA as indicated in the (−) lane.

senescent RNP complexes. The identity of genes was identical for both young and old cells. Scores were considered to be positive if the similarity score was more than 200 [15]. The genes identified were Homo sapiens *HDM2* gene (AF144029), intron 9 and exon 10, partial sequence; H. sapiens asthmatic clone 1 mRNA (AF095853), 3ʺ UTR; H. sapiens D15S1506 ca repeat region (AF018071), complete sequence, and H. sapiens partial *HR* gene for hairless protein (AJ277249). As these four human genes were the only gene candidates identified, we chose to score the sequences by the number of times a positive hit occurred in a (BLAST) similarity search for individually isolated clones. The sequence for *HDM2* that occurred most frequently was further analyzed to determine potential splicing enhancer/silencer elements by Zhang, et al. [16, 17] using the PESX utility (http://cubweb.biology.columbia.edu/pesx/). **Figure 3** shows a schematic illustration of the *HDM2* mRNA sequence that binds to hnRNP A1. The binding site is between exon 9 and intron 10. The regions marked in red are putative silencer sequences. We found that there were several matches to these sequences within the isolated *HDM2* mRNA from hnRNPA1 RNP complexes.

The identification of the human double minute 2 gene (*HDM2*) was of particular interest to us. It is the human homolog of the *mouse double minute 2 (MDM2)* and is a known oncogene [18–20]. It is a protein of ~90 kDa size and is usually localized in the nucleus of cells. Overexpression of HDM2 causes cells to proliferate uncontrollably as it facilitates the proteasomal degradation of p53 by acting as a ubiquitin ligase [21, 22].

The human murine double minute 2 (*HDM2*) gene is a 33-KB nucleotide sequence located on chromosome 12 (q14.3-q15). The gene consists of 12 exons and 11 introns [18, 20]. Transcription of the *HDM2* gene is controlled by two different promoters, referred to as P1 and P2 that are P53-independent and P53-dependent, respectively [19].

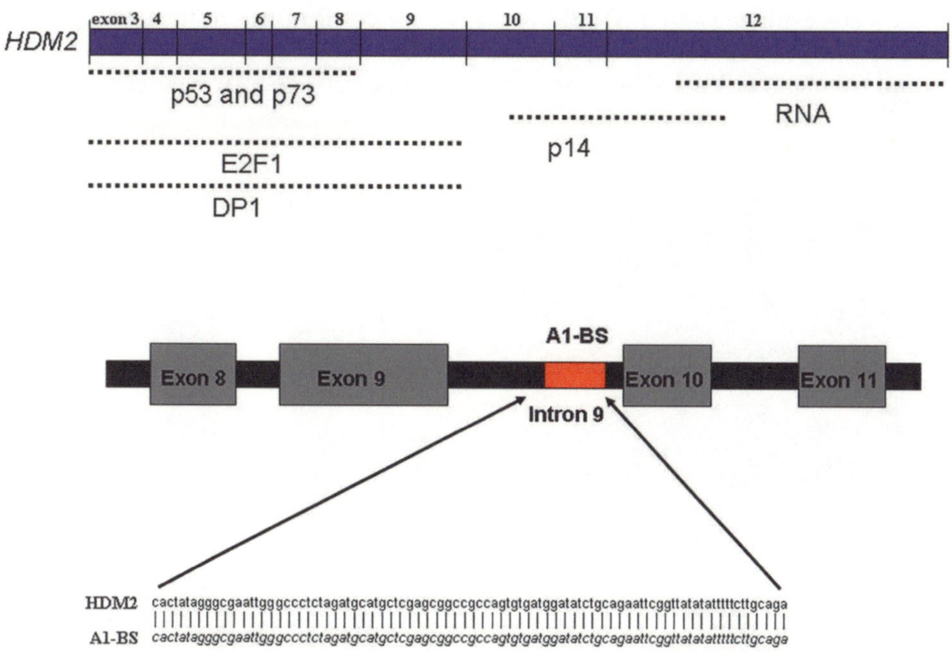

Figure 3.
PESX analysis. PESX (putative exon spliyancer/silencers) is a utility that can predict regions of an mRNA that may be involved in exon splicing to enhance the inclusion or exclusion of an exon. In addition to the identification of the putative splicing regulatory sequences, the sequences were also scanned for putative hnRNPA1-binding sites [UAG (G/a)] based on known binding sites.

The P1 promoter controls the basal expression of *HDM2* and is positioned upstream in the first exon of the *HDM2* gene [19, 20]. Transcription from the P2 promoter is highly regulated, responsible for the inducible expression of *HDM2*, and is found in the first intron [21]. The *HDM2* transcript is translated into a protein of 491 amino acids with multiple sizes ranging from about 50 to 90 kDa [22]. HDM2 can be regulated positively by AKT, a serine-threonine kinase, leading to the repression of p53 activity. AKT phosphorylates HDM2, which leads to its nuclear entry and subsequent attenuation of p53 activity and p53 degradation [23]. p14 (ARF), a protein product from the CDNK2A locus, is a negative regulator of *HDM2* [24]. The p14ARF protein binds to the central domain of HDM2, including the acidic region, leading to inhibition of the ability of HDM2 to act on p53 [25]·

To determine whether the full-length *HDM2* mRNA was bound to hnRNP A1 in RNP complexes, we isolated RNA from the complexes and used semi-quantitative to amplify full-length sequences. We found full-length *HDM2* in total RNA and complexes isolated from young cells (**Figure 4A**). To provide an estimation of the nature of the *HDM2* exons normally present in young fibroblasts, we amplified sequences from exon 1 to 2, exon 6 to 7, and exon 9 to 10. **Figure 4B** shows that of the regions amplified, the one that covered exon 9 to 10 was in the least abundance by a factor of 4 when compared with the most abundant region that spanned from exon 6 to 7. These results indicate that there may be preferential exon inclusion/exclusion in young fibroblasts in which hnRNP A1 is highly expressed.

2.4 Identification of p16INK4a mRNA in hnRNPA1 complexes

We have previously shown that changes in the expression of hnRNP A1 regulate the alternative splicing and mRNA levels of two mRNA isoforms of the INK4a locus known as p14^{Arf} and p16(INK4a) [10]. Both protein isoforms are growth suppressors and knockout of the INK4a gene allows cells to escape cellular senescence [11].

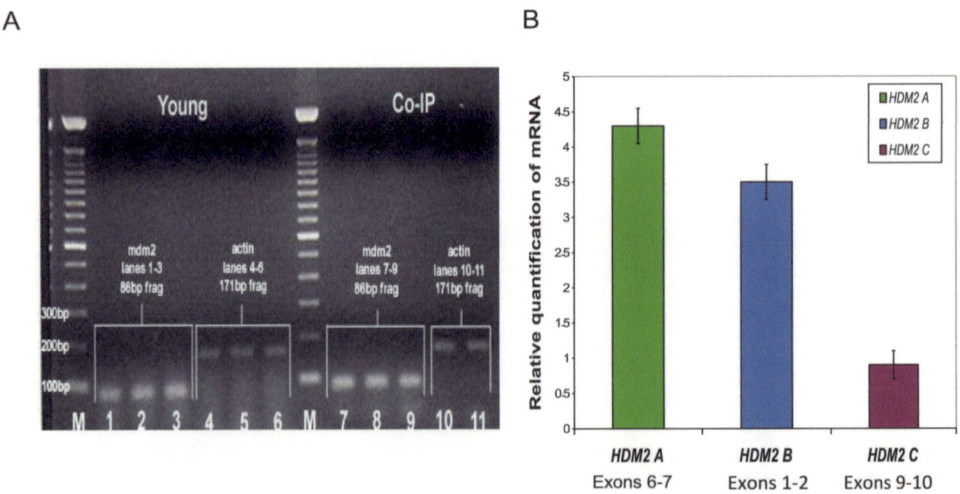

Figure 4.
HDM2 mRNA variant expression levels. RNA was isolated from co- immunoprecipitated hnRNPA1 protein complexes from young IMR90 fibroblasts and reverse transcribed. Three independent replicates of cDNA were used as templates for detecting HDM2 regions: Exons 6–7, HDM2A; exons 1–2, HDM2B and exons 9–10, HDM2C. Taqman primer-probe sets specific for these regions of the HDM2 gene were used in qRT-PCR assays and compared to actin mRNA levels.

Our previous studies have shown [9] that overexpression of hnRNA1 results in a preferential expression of the p14^{Arf} mRNA isoform, and an increase in the mRNA levels of both isoforms, thus suggesting a role for hnRNP A1 in control of cell proliferation and senescence [10]. In this study, we assessed the ability of hnRNP A1 to directly bind to INK4a transcripts in hnRNPA1 complexes. For this, we used AR5 cells and PRNS-1 (SV40-transformed clones of HS74 primary human bone marrow fibroblasts) since these cells express high levels of INK4a transcripts as compared with normal IMR-90 fibroblasts [26].

Figure 5 shows that the p16 transcript was amplified from hnRNPA1RNP complexes indicating that hnRNP A1 directly binds to p16 mRNA. We also measured the ability of hnRNP A1 to bind to actin mRNA as a positive control for our co-immunoprecipitation studies. Actin mRNA has been previously identified in hnRNPA1-RNP complexes [12]. **Figure 5** shows that in addition to p16, we were also able to detect actin mRNA in the RNP complexes.

2.5 Expression of *HDM2* is modulated by hnRNPA1 expression levels

We next sought to determine whether hnRNPA1 modulated the expression of *HDM2* mRNA levels. hnRNP A1 was overexpressed in young IMR-90 human fibroblast cells followed by real-time RT-PCR analysis using primer sets that amplified different regions of *HDM2* mRNA. We observed a significant decrease in the full-length *HDMd2* mRNA levels in cells overexpressing hnRNP A1 as compared with cells expressing the empty GFP Vector (**Figure 6B**). Downregulation of hnRNP A1

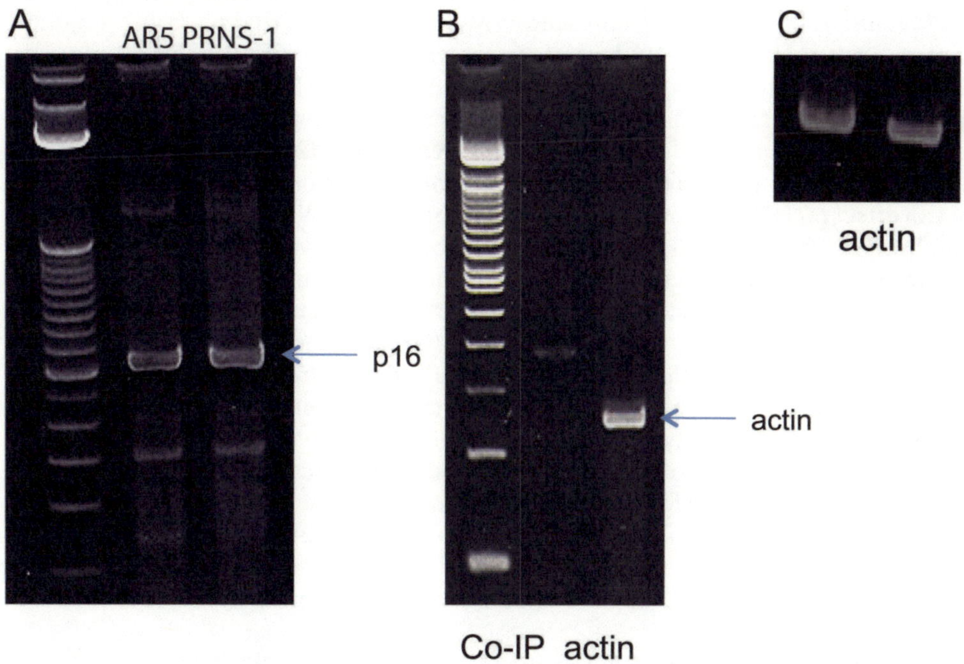

Figure 5.
Immunoprecipitation of specific mRNA hnRNP A1 RNP complexes. RNA was extracted from co-immunoprecipitated hnRNPA1 protein complexes isolated from AR5 and HS74-PRNS-1 cells and reverse transcribed. Primers set specific for p16 and actin were used to identify their respective mRNAs in hnRNP A1 RNP complexes. AR5 and HS74-PRNS-1 cells are SV40-transformed immortal cell lines.

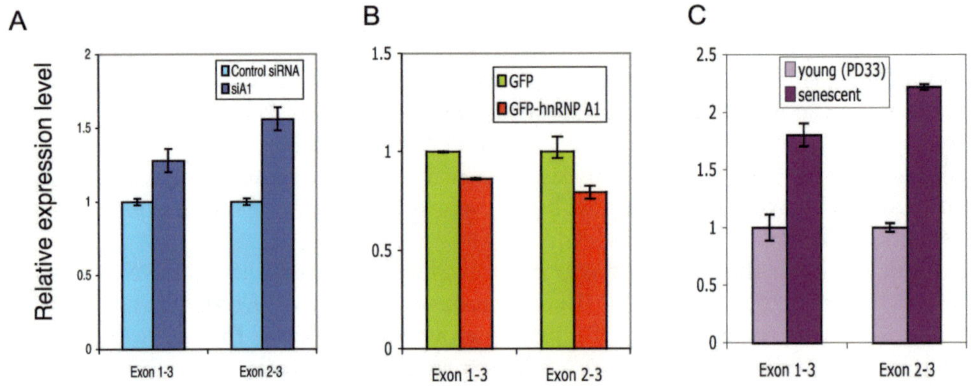

Figure 6.
Expression of HDM2 mRNA levels following alteration of hnRNP A1 expression. Panel a: Scrambled siRNA (control) or siA1 oligonucleotides were transfected into IMR-90 fibroblast cells. Real-time PCR was performed using primers for constitutive HDM2 mRNA (exon 1–3) and p53-inducible HDM2 mRNA (exon 2–3). RPLPO mRNA levels were used as an internal control. Panel B: Expression plasmids pEFGP empty vector or pEFGP-A1 were transfected into IMR-90 fibroblast cells. After 2 days of incubation, total RNA was extracted from cells and real-time PCR was performed with different sets of primers for detection of constitutive and p53-inducible HDM2 mRNA. RPLPO and GAPDH RNA levels were used as internal standards. Panel C: The steady-state endogenous levels of constitutive and p53-inducible HDM2 were measured in young and senescent mRNA.

by siRNA transfection showed increased levels of *HDM2* mRNA (**Figure 6A**). These results indicate that hnRNP A1 protein levels modulate the mRNA levels of *HDM2*. Since we had previously shown that hnRNP A1 expression and its subcellular distribution were altered during cellular senescence [27], we compared the endogenous *HDM2* mRNA levels in young and senescent cells. We found that there was a significant increase in *HDM2* mRNA levels in senescent cells as compared with young cells (**Figure 6C**). These results show that endogenous levels of HDM2 were consistent with our overexpression and siRNA results discussed earlier.

We also investigated whether the protein level of HDM2 was modulated by the level of hnRNP A1 protein expression. To determine whether endogenous hnRNP A1 has an effect on HDM2 protein expression, scrambled siRNA or siA1 was transfected into IMR-90 fibroblast cells. We found that upon siRNA knockdown of hnRNP A1, the protein level of HDM2 was not altered as shown in **Figure 7A**. hnRNP A2, which has overlapping biochemical activity with hnRNP A1, when inhibited by siRNA interference, did not affect HDM2 expression. On the other hand, overexpression of hnRNP A1 in young cells transfected with GFP-A1 resulted in a slight decrease of HDM2 protein levels and an increase in p53 levels when compared with cells transfected with the GFP-Empty vector (**Figure 7B**). The increase in p53 protein levels may be a result of the decreased HDM2 expression. A direct correlation between protein and mRNA levels for any given gene is complicated by varying processes. For instance, studies conducted by various groups such as Vogel et al. [28] pointed out that transcription, mRNA export, decay, translation, and protein degradation are key processes in determining steady-state protein concentration [28].

2.6 Identification of *HDM2* RNA sequences that bind to hnRNP A1

Our PESX analysis revealed that hnRNP A1 has a putative binding site within the intronic region between intron 9 and exon 10 of HDM2 (**Figure 4**). We obtained

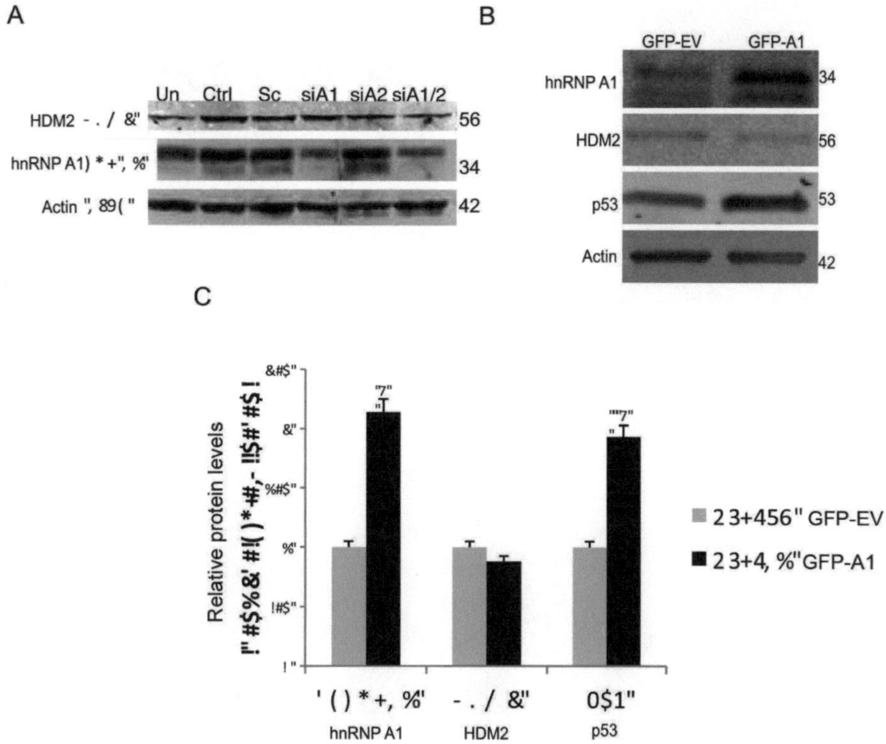

Figure 7.
*Effect of varying hnRNP A1 expression on HDM2 protein levels a. scrambled siRNA (control), siA1, siA2, or both oligonucleotides were transfected into IMR-90 fibroblast cells. About 30 µg of protein lysates were subjected to 12% SDS-PAGE and immunoblotted for hnRNPA1 and actin. The membranes were stripped and reprobed for total actin levels. B. IMR-90 cells were transiently transfected with GFP-hnRNP A1 and empty vector for 48 hours. Whole-cell lysates were prepared using RIPA lysis buffer. 30 µg of protein lysates were simultaneously subjected to 12% SDS-PAGE and immunoblotted for hnRNP A1, HDM2, p53 and actin. The membranes were stripped and reprobed for total actin levels. C. the relative protein levels was quantified by image density analysis software (image J), and the data were represented as mean value ± SEM (n = 3, and * $p < 0.05$ value).*

HDM2 constructs from Dr. Meek (University of Dundee). We performed biotin pull-down assay using MP4 construct that is similar to the MDM2-B isoform lacking p53-binding region followed by Western immunoblotting. We performed the biotin pull-down assay by first incubating the hnRNP A1 antibody (4B10) with Dynabeads Myone streptavidin. We also incubated the ^{Biotin}labeled *HDM2* mRNA (MP4 probe) or ^{Biotin}labeled *Β*-actin RNA probe with IMR-90 cell lysate and the mixture was incubated overnight. We performed Western blot analysis as described in Material and Methods to investigate the interaction between ^{Biotin}labeled *HDM2* mRNA and hnRNP A1. As shown in **Figure 8**, hnRNP A1 is directly associated with ^{Biotin}labeled *HDM2* mRNA (MP4 probe). The direct association between hnRNP A1 and ^{Biotin}labeled *HDM2* mRNA appears to be specific since these interactions were not observed with similar RNA-binding protein hnRNP A0. Furthermore, associations between hnRNP A1 and negative-control ^{Biotin}labeled *Β*-actin were not observed. These results demonstrate that hnRNP A1 has specific binding to ^{Biotin}labeled *HDM2* mRNA (MP4 probe). These results also show that hnRNP A1 can bind to the 3′ end of HDM2 as MP4 is the 3′ of HDM2 which is also a region devoid of HDM2 sequences upstream of exon 11.

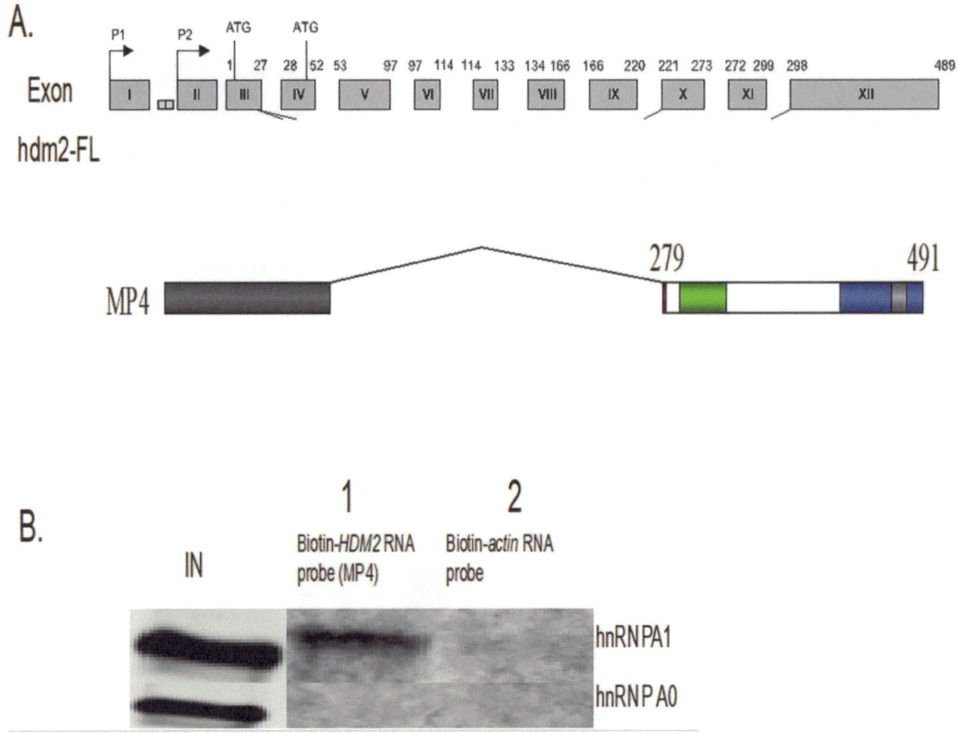

Figure 8.
HDM2 transcript (MP4) binds to hnRNP A1. A. Schematic representation of full-length HDM2 and mini-protein "MP" fragment of HDM2 MP4. B. the interaction between hnRNP A1 and biotinylated HDM2 mRNA (MP4) was examined by biotin pull-down assay. Biotinylated HDM2 and actin probes were incubated with IMR-90 cell lysates. Interactions were analyzed by Western blotting. The protein-RNA probe streptavidin complexes were subjected to electrophoresis on a 12% SDS-PAGE. The 4B10 monoclonal antibody and anti-actin antibody were used to detect hnRNP A1 and actin, respectively.

3. Discussion

The role of hnRNP A1 during cellular senescence is unclear. Significant alterations in its levels, localization, and activity in senescent cells suggest that hnRNP A1 may contribute to the senescent phenotype [27]. However, only a few gene targets are known for hnRNP A1 [12]. This prompted us to search for additional mRNA targets for hnRNP A1 in young and senescent IMR-90 cells. We used an RNA co-immunoprecipitation protocol [12] to identify mRNA new targets for the hnRNP A1 protein. We found that hnRNP A1 is bound to several mRNAs not previously identified. Of particular interest to us was the observation that hnRNP A1 bound to *HMD2* mRNA. Other RNA-binding proteins have been reported to bind to specific regions in *HDM2* mRNA. La antigen, an RNA-binding protein, was found to interact with HDM2 5′UTR in a BCR/ABL cell line resulting in increased HDM2 expression [29]. It was further demonstrated that translational regulation contributed to the increased HDM2 levels in BCR/ABL cells [29]. Nucleolin, a multifunctional nucleolar protein with defined roles in ribosomal RNA processing, has also been reported to bind to the NLS/NES and RING domain of HDM2 [30]. The expression of *HDM2* mRNA is transcriptionally regulated by p53 in response to stress such as DNA damage [31, 32]. We have found that the modulation of hnRNP A1 expression can regulate *HDM2* mRNA levels.

Posttranscriptional regulation of gene expression is important for the control of cellular processes such as cell proliferation, differentiation, development, and apoptosis [33]. RNA-binding proteins are the main regulators of post-transcriptional regulation [33]. hnRNP A1 is a multifunctional RNA-binding protein implicated in the regulation of major steps in posttranscriptional regulation of gene expression [14]. Upon observation that hnRNP A1 binds to *HDM2* mRNA, we sought to determine whether the modulation of the expression of hnRNP A1 had an effect on steady state *HDM2* mRNA levels. Given that our previous results demonstrated that overexpression of hnRNP A1 significantly decreased *HDM2* mRNA levels, we next asked whether the HDM2 protein levels were also lowered when hnRNP A1 was overexpressed. Our Western blot analysis data revealed that overexpression of GFP-A1 slightly decreased HDM2 protein levels (**Figure 7B**), whereas knockdown of hnRNP A1 did not have effect on HDM2 protein expression. We also observed that p53 levels increased upon GFP-A1 overexpression suggesting that the levels of p53 protein were increased by the decreased expression of HDM2 (**Figure 7B**). Previous studies have shown that HDM2 regulates p53 by targeting it for degradation [19]. Recent studies have demonstrated that overexpression of hnRNP A1 led to a reduction of *HDM2*-FL transcript levels in HaCat cells [34]. More importantly, it was also shown that UVB irradiation increased the binding of hnRNP A1 to *HDM2* pre-mRNA [34]. Our studies are consistent with these findings whereby we have found that overexpression of hnRNP A1 decreased *HDM2* transcript levels.

In this study, our findings suggest that hnRNP A1 binds to the ^{Biotin}labeled *HDM2* mRNA probe (MP4) that includes part of exons 11 and 12 of *HDM2* mRNA as shown in **Figure 8A**. Previous approaches as those used by S.J. Park et al. [35] demonstrated that hnRNP A1 associates with Drp1 mRNA at the 3′ UTR [35]. We applied this approach and found that hnRNP A1 binds to the *HDM2* MP4 mRNA and that this binding was specific as hnRNP A0 did not bind to the *HDM2* MP4 probe (**Figure 8B**). Our sequencing data of the HDM2 MP4 reveal that it contains a putative G/AGAAG nucleotide sequence similar to the 5′AGAAG 3′ high-affinity binding site found in the purine-rich 3′ splice site of c-src mRNA exon N1 [36]. hnRNP A1 has been shown to bind to these sites in c-H-ras and HIV TAT [37, 38]. Overall, our RNA-protein interaction experiments data strongly suggest that hnRNP A1 interacts with a region of *HDM2* transcript corresponding to its 3′ UTR. It has been previously reported that RNA-binding protein RNPC1 binds to the 3′UTR region in *HDM2* transcripts and inhibits its expression [39]. Thus, the *HDM2* 3′ UTR is bound by different RNA-binding proteins that might either repress or induce its expression. For example, HuR, an RNA-binding protein, has been shown to bind and stabilize *HDM2 via* its 3′ UTR. From our studies, we found that hnRNP A1 modulated the mRNA expression of HDM2. Therefore, the MP4 sequence maybe partially contributing to this modulation. Our findings are significant when taken in the context of RNA-binding protein contributing to the aging phenotype. Both HuR and hnRNPA1 are involved in regulating the senescent phenotype [40, 41]. Inhibiting HuR expression induces the senescent phenotype [41]. hnRNP A1 has been recently shown to antagonize cellular senescence through the SIRT1 pathway [42]. The research findings of the research project are important because they can add to the knowledge of the regulation of HDM2 gene expression.

4. Materials and methods

4.1 Cell culture and generation of senescent fibroblasts

The human lung fibroblast cell strain IMR90 from Coriell, NJ, was subcultured from early passage to terminal passage as previously described by Hubbard and Ozer [43] in Dulbecco's modified Eagle's medium and Ham's F10 medium in a 1:1 mixture supplemented with 10% fetal bovine serum. IMR90 fibroblasts at population doubling <35 were used in all experiments and are considered comparable to young fibroblasts as determined by gene expression profiles previously performed [43]. Senescent IMR90 in all experiments was at a population doubling of 62. For transfection experiments, once cells had reached 90% confluence, either the expression plasmid pEFGP (Control) or pEFGP-A1 was transfected into IMR-90 cells in DMEM/F10 media without FBS/penicillin using Lipofectamine 2000 (Invitrogen) and incubate at 37°C in a CO_2 incubator for 6 h.

4.2 RNA isolation and RNA-PCR to check for genomic contamination

After RNA was isolated as detailed above, to ensure that there was no genomic contamination, an RNA–PCR procedure was performed. 2 μL of template RNA was added to a PCR mixture using β-actin primers, which would detect genomic sequences if present. The total PCR reaction was composed of 2 μl of template RNA, 5 μl of 10× RT-PCR Buffer (100 mM Tris–HCl, pH 8.3, 500 mM KCl, 15 mM $MgCl_2$), 2.5 μl of dNTP mix (2.5 mM each dNTP), 0.25 μM each PCR primer, 0.5 unit of Thermostable DNA Polymerase (Novagen).

The primers for actin were: actin forward (5′CGCCGCCCTAGGCACCA3′) and actin reverse (5′TTGGCCTTAGGGTTCAGGGGGG 3′). For hnRNPA1, primer set were: hnRNP A1 forward (5′CTAAAGAGCCCGAACAGCTGAG 3′) and hnRNP A1 reverse (5′TCAGTGTCTTCTTTAATGCCACCA 3′). SYBR™ green stain. SYBR™ green can be visualized by blue fluorescence (Molecular Dynamics, Amersham) and quantified with ImageQuant software (Amersham).

5. Immunoblotting and protein analysis

Standard Western blotting protocols (Harlow et al. 1999) were to analyze specific proteins [44]. Protein extracts isolated from young and senescent fibroblasts were generated by washing cells were washed 3 times with 1X cold PBS and then, cultures were placed on ice. Cold RIPA (radioimmunoassay buffer containing NP-40 at 1%, sodium deoxycholate at 1%, sodium dodecyl sulfate at 0.1%, NaCl at 150 mM and Tris-HCl at 10 mM with protease inhibitors leupeptin at 0.1 μg/ml, pepstatin at 0.1 μg/ml, and phenylmethylsulfonyl fluoride at 1 mM) was added to culture dishes followed by scraping cells into cold microfuge tubes. The lysate was passed through a 21-gauge syringe needle to ensure complete lysis. Lysates were centrifuged at 10,000×g for 10 minutes at 4°C. The cleared lysate was collected and aliquots were prepared to estimate the amount of protein by the Bradford protein assay (Bio-Rad). Lysates were run on 8–12% acrylamide gels and then transferred in an electroblotting apparatus. Membranes (PVDF, Osmonics) were blocked with 5% non-fat milk in PBS.

Monoclonal antibody 4B10 (1:10,000) was used to detect hnRNPA1/A1. An anti-actin monoclonal antibody (1:5000) was obtained from Chemicon. Antibodies specific for HDM2 were graciously provided by Dr. Jill Bargonetti. Secondary antibody, goat IgG, or mouse IgG conjugated with HRP were used for visualization of bands using the ECL kit (Amersham).

6. Overexpression of hnRNP A1 by transient transfection

Young IMR-90 cells were cultured in 10-cm plates. After approximately 24 h of incubation when the cells reached 90% confluence, the expression plasmid pEFGP (control) or pEFGP-A1 was transfected into IMR-90 cells in DMEM/F10 media without FBS/penicillin using Lipofectamine 2000 (Invitrogen) and incubated at 37°C in CO_2 incubator for 6 h. We changed the media to DMEM with FBS and without penicillin and incubated for 48 h at 37°C.

7. RNA co-immunoprecipitation protocol

The RNA co-immunoprecipitation protocol was a modified version published by Mili et al. [12] that included a short immunoprecipitation step that minimized degradation of protein-associated RNA.

8. Confirmation of gene expression using real-time PCR

Real-time PCR experiments on selected genes were performed using an Applied Biosystems 7500 real-time PCR system that utilizes TaqMan gene expression assays for the following genes: mdm2; Human GAPD (GAPDH) Endogenous Control FAM/MGB (4333764F). Reactions were performed according to standard methods using the universal 10X PCR TaqMan mix, at a final reaction volume of 25 μL (Applied Biosystems).

9. Cloning protocol and sequencing

PCR products were ligated into the pCR 2.1 (Invitrogen, TA cloning kit) cloning vector that utilizes the single dT overhangs that are a by-product of PCR reactions catalyzed by Taq polymerase. The ligation reaction was performed at 14°C overnight using T4 DNA ligase and 3 μL of fresh PCR product (Invitrogen protocols).

10. Sequence analysis using BLAST and PESX

Vector sequences were subtracted from the sequences obtained. The rest of the sequence was compared against known sequences using the BLAST tool (www.ncbi.nlm.nih.gov). Sequences were chosen based on being previously identified as human genes and either in the coding regions or in the flanking regions of the mRNA sequences of known genes.

11. Biotin pull-down assay

Biotinylated transcripts were obtained by reverse transcription with the Maxiscript Kit (Invitrogen) according to manufacturer instructions and as previously described above in Section 3.2. The biotin pull-down assay was performed by first incubating hnRNP A1 antibody (4B10) with Dynabeads Myone streptavidin (Invitrogen) for 1 hour at 4 C. Also, incubated ^{Biotin}labeled *HDM2* mRNA (MP4 probe) or B-actin RNA probe with 25 μg of IMR-90 cell lysate for 1 h at RT. Following this incubation, we added the biotinylated RNA probes and protein lysate mixtures to the Myone streptavidin beads coated with 4B10 (hnRNP A1 antibody) and performed a second overnight incubation, immediately subjected the protein-RNA complexes to Western blot analysis as described in Section 2.4 to detect specific proteins bound to biotinylated transcripts.

Acknowledgements

We are grateful to Alisa Chalmers for technical assistance. This research was supported by NIH/NCI U54CA137788/U54CA132378 to KH and to NIH/NIMHD 3G12MD007603-30S2 and NIH RISE R25GM056833 to HM. We also thank Dr. Serafin Pinol-Roma for hnRNP specific antibodies and consultation on the RNA co-immunoprecipitation assays. We also thank Dr. Bargonetti for the HDM2 antibody and Dr. Meek for providing HDM2 plasmid constructs. Heriberto Moran and Shanaz A. Ghandhi contributed equally to this work.

Conflict of interest and financial disclosures

There are no conflicts of interest nor financial interest or benefits.

Abbreviations

HDM2	Human Double Minute 2
UTR	Untranslated region
hnRNPA1	Heterogeneous nuclear ribonucleoprotein A1
RNP	Ribonucleoprotein
MP4	Biotin-labeled HDM2 RNA probe
mRNP	Messenger ribonucleoprotein

Author details

Heriberto Moran[1,2,*,†], Shanaz A. Ghandhi[1,2,†], Naoko Shimada[1] and Karen Hubbard[1,2]

1 The City College of New York, City University of New York, NY, USA

2 The Graduate Center of CUNY, New York, NY, USA

*Address all correspondence to: hmoran@ccny.cuny.edu

† These authors contributed equally to this work.

References

[1] Cristofalo VJ, Volker C, Francis MK, Tresini M. Age-dependent modification of gene expression in human fibroblasts. Critical Reviewes in Eukaryotic Gene Expression. 1998;**8**:43-80

[2] Francis R, Campisi J. Four faces of cellular senescence. The Journal of Cell Biology. 2011;**192**(4):547-556

[3] Wei W, Hemmer MR, Sedivy MJ. Role of p14ARF in replicative and induced senescence of human fibroblasts. Molecular and Cellular Biology. 2001;**21**(20):6748-6757

[4] Mayena A, Krainer AR. Regulation of alternative pre-mRNA splicing by hnRNPA1 and splicing factors SF2. Cell. 1992;**68**:365-375

[5] Burd CG, Dreyfuss G. RNA binding specificity of hnRNP A1: Significance of hnRNP A1 high-affinity binding sites in pre-mRNA splicing. The EMBO Journal. 1994;**13**(5):1197204

[6] Dreyfuss G, Matunis M, Pinol-Roma S, Burg CG. hnRNP proteins and the biogenesis of mRNA. Annual Review of Biochemistry. 1993;**62**:289-321

[7] Dreyfuss G. Structure and function of nuclear and cytoplasmic ribonucleoprotein particles. Annul Reviews Cell Biology. 1986;**2**:459-498

[8] Bai Y, Lee D, Yu T, Chasin LA. Control of 3′ splice site choice in vivo By ASF/SF2 and hnRNPA1. Nucleic Acids Research. 1999;**27**(4):1126-1134

[9] Eperon IC, Makarova OV, Mayeda A, Munroe SH, Caceres JF, Hayward DG, et al. Selection of alternative 5'splice sites: Role of U1 snRNP and models for the antagonistic effects of SF2/ASF and hnRNPA1. Molecular and Cell Biology. 2000;**20**:8303-8318

[10] Zhu D, Xu G, Ghandi SA, Hubbard K. Modulation of the expression of P16INK4a and P14ARF by hnRNP A1 and A2 RNA binding proteins: Implications for cellular senescence. Journal of Cellular Physiology. 2002;**193**(1):19-25

[11] Dimri GP, Hahana K, Acosta M, Campisi J. Regulation of a senescence checkpoint response by the E2F1 transcription factor and P14ARF tumor suppressor. Molecular and Cellular Biology. 2000;**20**(1):273-285

[12] Mili S, Shu HJ, Zhao Y, Pinol-Roma S. Distinct RNP complexes of shuttling hnRNP proteins with pre-mRNA and mRNA: Candidate intermediates in formation and export of mRNA. Molecular and Cellular Biology. 2001;**21**(21):7307-7319

[13] Allemand E, Guil S, Myers M, Moscat J, Caceres JF, Krainer AR. Regulation of heterogenous nuclear ribonucleoprotein A1 transport by phosphorylation in cells stressed by osmotic shock. PNAS. 2005;**102**(10): 3605-3610

[14] Jean-Phillippe J, Paz S, Caputi M. hnRNP A1: The Swiss army knife of gene expression. International Journal of Molecular Sciences. 2013;**14**(9): 18999-19024

[15] Altschul SF, Gish W, Miller W, Myers EW, Lipman DJ. Basic local alignment search tool. Journal of Molecular Biology. 1990;**215**(3):403-410

[16] Zhang XH, Kangsamaksin T, Chao MS, Banerjee JK, Chasin LA.

Exon 17 inclusion is dependent on predictable exonic splicing enhancers. Molecular and Cellular Biology. 2005;**25**(16):7323-7321

[17] Zhang XH, Chasin LA. Computational definition of sequence motifs governing constitutive exon splicing. Genes & Development. 2004; **18**(11):1241-1250

[18] Bartel F, Taubert H, Harris LC. Alternative and aberrant splicing of MDM2 mRNA in human cancer. Cancer Cell. 2002;**2**(1):9-1

[19] Barak Y, Juven T, Haffner R, Oren M. Mdm2 expression is induced by wild type p53 activity. The EMBO Journal. 1993;**12**(2):461-468

[20] Iwakuma T, Lozano G. MDM2, an introduction. Molecular Cancer Research. 2003;**1**(14):993-1000

[21] Manfredi JJ. The Mdm2-p53 relationship evolves: Mdm2 swings both ways as an oncogene and a tumor suppressor. Genes & Development. 2010;**24**:1580-1589

[22] Olson DC, Marechal V, Momand J, Chen J, Romocki C, Levine AJ. Identification and characterization of multiple mdm-2 proteins and mdm-2-p53 protein complexes. Oncogene. 1993;**8**(9):2353-2360

[23] Mayo LD, Donner DB. A phosphatidylinositol 3-kinase/AKt pathway promotes translocation of Mdm2 from the cytoplasm to the nucleus. PNAS. 2001;**98**(20): 11598-11603

[24] Honda R, Yasuda H. Association of p19(ARF) with Mdm2 inhibits ubiquitin ligase activity of Mdm2 for tumor suppressor p53. The EMBO Journal. 1999;**18**(1):22-27

[25] Sherr CJ. Divorcing ARF and p53: An unsettled case. Nature Reviews. Cancer. 2006;**6**:663-673

[26] Neufeld SD, Ripley S, Henderson A, Ozer LH. Immortalization of human fibroblasts transformed by origin-defective simian virus 40. Molecular and Cell Biology. 1987;**7**(8):2794-2802

[27] Shimada N, Rios I, Moran H, Sayers B, Hubbard K. p38 MAP kinase-dependent regulation of the expression level and subcellular distribution of heterogenous nuclear ribonucleoprotein A1 and its involvement in cellular senescence in normal human fibroblasts. RNA Biology. 2009;**6**:293-304

[28] Vogel C, Abreu RS, Ko D, Le SY, Shapiro BA, Burns SC, et al. Sequence signatures and mRNA concentration can explain two-thirds of protein abundance variation in a human cell line. Molecular Systems Biology. 2010;**24**(6):1-9

[29] Trotta R, Vignudelli T, Candini O, Intine VR, Pecorari L, Guerzoni C, et al. BCR/ABL activates mdm2 mRNA translation via the La antigen. Cancer Cell. 2003;**3**:145-160

[30] Bhatt P, d'Avout C, Kane NS, Borowiec JA, Saxena A. Specific domains of nucleolin interact with Hdm2 and antagonize Hdm2-mediated p53 ubiquitination. The FEBS Journal. 2012;**279**(3):370-383

[31] Chang C, Freeman JD, Wu H. PTEN regulates Mdm2 expression through the P1 promoter. The Journal of Biological Chemistry. 2004;**279**(28):29841-22984

[32] Riley MF, Lozano G. The many faces of MDM2 Binding partners. Genes & Cancer. 2012;**3**(3-4):226-239

[33] Glisovic T, Bachorik JL, Yong J, Dreyfuss G. RNA-binding proteins and

post-transcriptional gene. FEBS Letters. 2008;**582**(14):1977-1986

[34] Feng J, Li J, Tong L, Tan L, Wu S. The involvement of splicing Factor hnRNP A1 in UVB-induced alternative splicing of hdm2. Photochemistry and Photobiology. 2016;**92**:318-324

[35] Park SJ, Lee H. Jo YK, Shin JH, Kim, HB, Seo HM, Rubinsztein D., Koh JY, Lee EK, Cho DH. Heterogenous nuclear ribonucleoprotein A1 post-transcriptionally regulates Drp1 expression in neuroblastoma cells. Biochimica et Biophysica Acta 2015;1829(12):1423-1431.

[36] Rooke N, Markovtsov V, Cagavi E, Black DL. Roles for SR proteins and hnRNP A1 in the regulation of c-src exon N1. Molecular and Cellular Biology. 2003;**23**(6):1874-1884

[37] Damgaard CK, Tange TO, Kjems J. hnRNP A1 controls HIV-1 mRNA splicing through cooperative binding to intron and exon splicing silencers in the context of a conserved secondary structure. RNA. 2002;**8**(11):1401-1415

[38] Guil S, Gattoni R, Carrascal M, Abian J, Stevenin J, Bach-Elias M. Roles of hnRNPA1, SR proteins, and p68 helicase in c-H-ras alternative splicing regulation. Molecular and Cellular Biology. 2003;**23**(8):2927-2941

[39] Xu E, Zhang J, Chen X. MDM2 expression is repressed by the RNA binding protein RNPC1 via mRNA stability. Oncogene. 2013;**32**(17): 2169-2178

[40] Lee BP, Pilling LC, Emond F, Flurkey K, Harrison DE, Yuan R, et al. Changes in the expression of splicing factor transcripts and variations in alternative splicing are associated with lifespan in mice and humans. Aging Cell. 2016;**15**(5):903-913

[41] Hashimoto M, Tsugawa T, Kawagishi H, Asai A, Sugimoto M. Loss of HuR leads to senescence-like cytokine induction in rodent fibroblasts by activating NF-κB. Biochimica et Biophysica Acta. 2014;**1840**(10): 3079-3087

[42] Wang H, Han L, Zhao G, Shen H, Wang P, Sun Z, et al. hnRNP A1 antagonizes cellular senescence and senescence-associated secretory phenotype via regulation of SIRT1 mRNA stability. Aging Cell. 2016; **15**(6):1063-1073

[43] Hubbard K, Ozer HL. Chapter 12; Senescence and immortalization of human cells. In: Studzinski G, editor. Cell Growth and Apoptosis: A Practical Approach. England: IRL Press; 1995. pp. 229-249

[44] Harlow E, Lane D. Using Antibodys: A Laboratory Manual. New York, NY: Cold Spring Harbor Laboratory Press; 1999

Chapter 3

Genetic and Epigenetic Influences on Cutaneous Cellular Senescence

Tapash Jay Sarkar, Maiko Hermsmeier, Jessica L. Ross and G. Scott Herron

Abstract

Skin is the largest human organ system, and its protective function is critical to survival. The epithelial, dermal, and subcutaneous compartments are heterogeneous mixtures of cell types, yet they all display age-related skin dysfunction through the accumulation of an altered phenotypic cellular state called senescence. Cellular senescence is triggered by complex and dynamic genetic and epigenetic processes. A senescence steady state is achieved in different cell types under various and overlapping conditions of chronological age, toxic injury, oxidative stress, replicative exhaustion, DNA damage, metabolic dysfunction, and chromosomal structural changes. These inputs lead to outputs of cell-cycle withdrawal and the appearance of a senescence-associated secretory phenotype, both of which accumulate as tissue pathology observed clinically in aged skin. This review details the influence of genetic and epigenetic factors that converge on normal cutaneous cellular processes to create the senescent state, thereby dictating the response of the skin to the forces of both intrinsic and extrinsic aging. From this work, it is clear that no single biomarker or process leads to senescence, but that it is a convergence of factors resulting in an overt aging phenotype.

Keywords: skin, DNA damage, telomeres, epigenetics, immunosenescence, inflammaging

1. Introduction

A consensus agreement on the definition of cellular senescence may be stated as a viable but non-proliferative condition distinct from the G0 quiescent phase of the cell cycle and postmitotic terminal differentiation [1, 2]. While aged living organisms accumulate senescent cells, aging and senescence are not synonymous terms—the cellular and molecular pathways that eventuate in the senescent state can be activated by diverse mechanisms, not necessarily chronologic aging nor the limit of replicative cell division. It was the latter phenomenon, in fact, that led early investigators to the original concept of cellular senescence as an *in vitro* observation; that replicating fetal "skin tissue cells" stop dividing at a certain passage number, the so-called "Hayflick Limit" [3]. This review focuses on our current understanding of how cellular senescence occurs in the skin, its irreversible (and possibly reversible) characteristics, description of known trigger points involving genetic and epigenetic factors and their clinical implications in health and disease.

Cellular senescence is characterized by cell cycle arrest [4], the expression of senescence associated secretory phenotype (SASP) [5, 6], damage to DNA [7–9], deregulated metabolic profile [2, 10], changes to the epigenome [11] and transcriptome [12], resistance to apoptosis [13, 14], and altered immune surveillance [15, 16]. It can be triggered by multiple factors [2], the mechanisms of which appear to categorize the 'type' of senescence into two main groups; so-called replicative senescence (RS) due to shortened telomeric DNA resulting from excessive cell division cycles [17–20]; and a state generally termed 'premature senescence' (PS), in which both oncogene-induced senescence (OIS), triggered by activation of oncogenes such as *ras* [21], and several other 'molecular stresses' [4] also eventuates in the senescent phenotype.

There are a variety of biomarkers for cellular senescence but not all senescent cells express the same biomarkers due to these differential molecular induction pathways. Several senescent biomarkers have been identified in the skin [22]; however, it is currently unclear how the multitude of cell types that comprise this tissue respond to senescence-inducing triggers and how this correlates with skin aging, other than the fact that senescent cells accumulate in all skin compartments with age, just like other organ systems. What is becoming clear, however, is that cellular senescence plays critical roles in the pathobiology of skin aging and disease [23].

2. Aging of the skin

Both intrinsic (time, genetic and hormonal) and extrinsic (environmental) factors contribute to skin aging. Old skin not only appears clinically different from young skin but has altered physiology due to a combination of molecular, cellular, and biochemical processes, and tracing the pathogenic origin of the 'skin aging phenotype' remains a work in progress. From a clinical perspective however, the skin of most people older than 6–7 decades of life, particularly in photo-exposed areas, is thinner, looser, less tethered to underlying tissue, more wrinkled, more translucent with more visible capillary vessels, more discolored, drier, and less padded by the subcutaneous layer [24, 25]. Scalp skin also ages, commonly observed as pigment loss (graying), and most people experience hair loss as another inevitable esthetic problem.

Anatomically, the structure of human integument tissue we call 'skin' is composed of ectodermal-derived epithelial cells layered as stratified squamous epithelium on top of mesenchymal-derived dermis separated by a specialized basement membrane zone (BMZ) called the dermal-epidermal junction (DEJ). Directly below the dermis is the fatty hypodermis (or subcutaneous layer) separating fascia and muscle from the skin. Epithelial-mesenchymal interactions that occur during embryogenesis (and wound repair) contribute to the formation of glandular structures buried within these compartments (called adnexa) which are comprised of eccrine, apocrine, sebaceous and hair follicle structures. Peripheral nerves and blood vessels traverse the subcutaneous and dermal layers and together with all other structures of the integument, serve the functions of barrier protection, retention of heat and water, sensation, contractility, and lubrication [26].

This multi-compartmental system is the largest organ of the body and is composed of about 20 different cell types responsible for skin function and its stratification [27], all of which also change with age to contribute to the overall 'skin aging phenotype'. At the microscopic level, skin tissues of older individuals exhibit common characteristics regardless of whether sun protected or chronically exposed to ultraviolet (UV) light [28]. The most obvious and well documented structural changes include epidermal thinning, loss of rete ridges and flattening of dermal papillae [29, 30], keratinocyte and melanocyte architectural changes [31, 32], BMZ/DEJ alterations [33], less dense and altered reticular dermal collagen structure [34], accumulation of altered elastin and elastic fiber structural abnormalities [35], altered shape and loss of papillary dermal capillary loops [36–38], and size and structural alterations in glandular structures of the eccrine, apocrine and sebaceous units [39]. Concomitant with morphologic changes observed in aged skin, senescent cell populations increase in all skin compartments.

2.1 Effects of aging on epidermal structure and function

The epidermis, consisting of 5 different layers of keratinocytes, continuously renews itself on an approximate 27-day cycle by a differentiation program involving basal cells which are maintained and replenished by stem cells residing in the bulge region of the hair follicle and the interfollicular epidermis [40]. Of particular importance to aging of the epidermal compartment is the general concept that the cellular microenvironment (or niche) of stem cell populations plays a critical role in homeostatic resupply of transient amplifying basal cells [41]. The epidermis maintains a dynamic equilibrium by proliferating in the basal layer that is attached to the DEJ, then cell division ceases and basal keratinocytes undergo terminal differentiation while spatially migrating towards the top of the epidermis. During this transition, keratinocytes acquire specialized cytoskeletal components and create an intercellular diffusion barrier, eventually forming the outermost epidermal layer called the stratum corneum (SC). The SC is a specialized acidic, hydrophobic, protein-lipid-carbohydrate flexible 'shell' resistant to wear and tear, water loss, and invasion of microbes [42]. The "barrier function" of the skin is derived from the SC.

The epidermal compartment appears to deal with the ravages of extrinsic aging in a fundamentally different way than the dermal compartment because terminally differentiated (cell cycle arrested) keratinocytes are continuously shed, thus removing accumulated DNA and other macromolecular damage that otherwise trigger the senescent phenotype. But since the epidermis is continuously replenished by stem cells arising from the interfollicular niche, its alteration can affect epidermal biology in profound ways. In fact, epidermal stem cell niche can be affected during aging by both basal keratinocytes [43] and dermal fibroblasts [44]. Niche microenvironments can be altered by intrinsic and extrinsic aging at cell-cell, cell-matrix and paracrine signaling levels, leading to stem cell depletion and the 'atrophic epidermal phenotype' observed in intrinsic aged skin [45].

Many other cell types localize to the epidermis, including pigment-producing melanocytes found in the basal layer that protect against UV radiation. Pigment is synthesized within the melanocyte but transferred to neighboring basal keratinocytes (and specialized hair follicle-associated keratinocytes) via a complex melanosomal exo/phagocytosis mechanism localizing at the dendritic tips of melanocytes which interdigitate with up to 20 keratinocytes [46]. Melanocyte dysfunction associated with extrinsic aging (mostly photoaging) manifests clinically as abnormally dispersed and/or diminished melanin pigment (i.e., dyschromia, lentigines, and in the scalp, canities). Senescence of the melanocyte has been observed both *in vitro* and *in vivo* and the molecular pathways involved identified [47]. In fact, based on biomarker (e.g., P16) expression, senescent melanocytes appear to represent most senescent cells in aged epidermis [48] and their contributions to development of the epidermal atrophic phenotype via autocrine and paracrine (i.e., SASP) mechanisms have been identified [49].

The epidermal immune system is a network of resident antigen-presenting dendritic Langerhans cells (LC) thought to function as immune sentinels [50] together with trafficking lymphoid immune cells including resident memory T cells as CD8+

and CD4+ cells [51]. Of interest, a specialized CD4 + T cell (Treg) residing near the hair follicle bulge areas (located in the dermis) has been shown to play a role in hair growth cycling [52]. Skin aging is associated with variable deterioration of both adaptive and innate immune function, generally referred to as cutaneous 'immunosenescence'. This term has become controversial in the literature [16] because immune cell senescence is, in part, a physiologic adaptive response to survival and fitness of the organism. Its use to describe altered skin immune responses with age appears appropriate in the context of inflammaging [53], since the concept of immunosenescence encompasses both systemic chronic, low-grade inflammation [i.e., elevated serum levels of interleukins IL-6 and IL-8 and increased tumor necrosis factor alpha (TNF-α), etc.] and the presence of dysfunctional immune responses in various skin compartments apparently related to both tissue level RS and PS. Currently unknown, however, is whether cutaneous inflammaging is a cause or an effect of dysfunctional innate immune responses observed in the elderly and whether cellular senescence is responsible.

Examples of cutaneous aged-related immune dysfunction include reported reductions in the number and functionality of LC in aged skin and this correlates with both age related defective epidermal barrier function and inflammaging [54]. Likewise, defective physiologic immune clearance of senescent cells that contribute to aged skin pathologies have been demonstrated in dermal fibroblasts by the observation that these cells express a nonclassical major histocompatibility antigen (HLA-E). Its increased expression appears to block activation of natural killer (NK) cells and CD8+ cells responsible for clearing damaged cells, suggesting that evasion of dermal immunosurveillance leads to persistence of senescent dermal fibroblasts [55].

Aging is a clinical comorbidity in many skin diseases pathogenically linked to defective cutaneous innate and adaptive immune responses [53]. The incidence and prevalence of autoimmune blistering disorders such as bullous pemphigoid (BP), pemphigus vulgaris, and epidermolysis bullosa acquisita are all increased in older populations, BP being the most common example [56]. Likewise, aging is a comorbidity in the development of skin cancers, and the loss of immunosurveillance due to dysfunctional LCs is thought to contribute to progression of both non-melanoma skin cancer (NMSC) [57] and melanoma [58].

Another unique cell type scattered along the DEJ, considered part of the epidermal compartment, and possessing mixed neuronal, endocrine, and immunologic functions (as well as embryonic origin) is the Merkel cell (MC) [59]. Its involvement in the skin's somatosensory system is key to the sense of fine touch discrimination, which is decreased in the elderly [60]. In glabrous skin MC form complexes with intraepidermal sensory neurites found at the DEJ termed 'touch domes' or Merkel's discs. Digital skin of aged humans contains less of these complexes, lower density of MC and decreased expression of the stretch-activated ion-channel component Piezo2 [61]. Occurring mainly in aged humans, a rare but very aggressive skin cancer, Merkel cell carcinoma (MCC) has attracted recent attention due to its mysterious etiopathogenesis. 80% of MCC is associated with integration of a newly identified polyoma virus (MCPyV) [62], whereas 20% appear linked to accumulation of UV light-induced somatic mutations [63]. As detailed in the next section on epigenetics, it is of interest that the majority of MCC display expected chronologic age but DNA methylation patterns of epigenetically youthful cells [64].

2.2 Aging and the dermis

The dermal compartment is divided into superficial, reticular, and deep dermis with unique cellular, vascular, extracellular matrix (ECM), and adnexal components that define each space. Much of the 'business-end' of the dermis is localized to the superficial dermal compartment and the DEJ is central to its structure and functionality. It is considered part of both the epidermis and the superficial dermis because cellular components of each layer contribute to its synthesis, maintenance, and renewal. Serving as an adhesive scaffolding for basal keratinocytes, a shear-resistant Velcro-like surface securing the dermis to the epidermis, a complex paracrine factor-sequestering and mechano-transducer signaling layer, the DEJ modulates a remarkable number of cutaneous cellular processes involved in skin structure, function, regeneration, and resistance to trauma [33]. Comprised mainly of Type IV collagen and laminin, like other BMZs, DEJ complexity has been dissected at the molecular level to reveal a complex network of other collagens (VII, XVII and XVIII), 4 different isoforms of laminin (511, 521, 311 and 332), perlecan, nidogens, SPARC, fibrulins-1 and -2, dystroglycans, and integrins a3b1 and a6b4. All of these DEJ components are altered during aging and these changes correlate with age dependent increases in both DEJ thickness and stiffness [65].

Immediately beneath the DEJ, forming nipple-like structures projecting into the epidermal compartment and containing unique ECM, microvasculature, specialized fibroblasts, and dermal mesenchymal stem cells is the papillary dermis (PD). Here, undulating dermal protrusions interdigitate with epidermal rete ridges (pegs) to increase surface area for nutrient transfer, trafficking of immune cells, and increased tensile strength. The transition of superficial to reticular dermis is static and defined mostly by changes in ECM structure but dynamic during repair and disease. The majority of space in reticular dermis is occupied by thick bundles of interstitial collagens I and III, elastin and fibrillin fibers, and amorphous 'ground substance' comprised of hyaluronic acid, proteoglycans and glycoproteins. Like the DEJ, most if not all these dermal ECM components are altered and/or dysfunctional due to aging [35]. Both intrinsic and extrinsic aging correlate with the loss of rete ridges and flattening of dermal papillae [29, 30]. Compared to young, non-exposed and old photo-protected skin, the PD of chronically photodamaged skin displays marked structural changes, the most dramatic feature of which is the presence of "solar elastosis" in the superficial dermis. Solar elastosis consists of pathologically altered elastin fibrils [66] that present as dense accumulations of amorphous material best visualized with trichrome staining.

The cellular composition of these dermal sub-compartments is a complex mix of fibroblasts, endothelial cells, myofibroblasts, macrophages, mast cells, trafficking immune cells, adipocytes, various stem cells, sensory neurites, and the differentiated cellular components of dermal adnexa (including the hair follicle). An example of such cellular complexity, the significance of which continues to evolve, is the apparent post-natal plasticity of the dermal fibroblast. Single cell RNA sequencing has revealed at least four different subpopulations of human dermal fibroblasts [67, 68], and skin aging has been demonstrated to have a strong effect on both dermal 'fibroblast' phenotype and functionality. For example, young papillary dermal fibroblasts can direct reformation of youthful DEJ and epidermal structure and function, whereas old papillary and/or reticular dermal fibroblast populations cannot [69]. Furthermore, it is the senescent PD fibroblast and CD271+, laminin 332-expressing interfollicular stem cells that contribute to age-associated pathologic remodeling of the DEJ [33, 70]. Recent attention has focused on specific dermal fibroblast subpopulations and their involvement in wound healing, fibrosis, and loss of epidermal stem cell 'stemness' due to niche signaling dysfunction. The homeobox gene engrailed-1 (*EN-1*) expression appears to distinguish two types of fibroblasts; those cells expressing *EN-1* are associated with fibrotic healing phenotype whereas *EN-1* negative fibroblasts promote physiologic remodeling [71, 72]. Epigenetic modulation of the fibrotic phenotype is reviewed in the next section.

2.3 Subcutaneous layer involvement in skin aging dysfunction

The subcutaneous compartment (hypodermis) is composed mainly of cellular lipid storage units (adipocytes) separated by thin weblike networks of specialized ECM stroma containing microvasculature, adipose derived mesenchymal stem cells (ADSC) and immune cells. It functions as a thermoregulatory and shock-resistant barrier, as well as a reservoir of bioactive factors involved in systemic lipid metabolism, energy balance, and endocrine function [73]. Subcutaneous fat also undergoes age-related changes that are generally like the epidermal and dermal compartments where an 'atrophic' phenotype becomes clinically evident. With aging, subcutaneous fat deposits in various body locations disappear and/or are redistributed to visceral locations elsewhere in the body, causing esthetic concerns; this redistribution is associated with a variety of systemic age-related disease states, including insulin resistance, metabolic syndrome, cardiovascular disease, and obesity [74, 75]. Of note, senescent cells have been shown to accumulate in aged adipose tissue [76], contributing to systemic inflammaging. Experimental clearance of senescent cells can dramatically affect the redistribution of fat from the visceral to the subcutaneous compartment and decrease SASP expression [77]. The mechanism(s) of adipose cell senescence has not been clearly defined; however, ADSC exhaustion, oxidative stress by reactive oxygen species (ROS), and niche disruption appear to play important roles [78].

The influence of adipogenic hormones in skin aging and senescence has received recent attention with the discovery that UVB-light induced PS in human keratinocytes can be rescued by adiponectin via its suppression of inflammatory signaling pathways and human beta defensin-2 (*hBD2*) expression [79]. Human dermal fibroblasts express adipokine receptors and both leptin and adiponectin have been shown to stimulate expression of the ECM components hyaluronic acid and interstitial collagen [80]. These adipogenic hormones secreted by subcutaneous fat cells thus appear to represent paracrine cutaneous anti-aging factors for both the epidermal and dermal compartments.

'Fat grafting' has become a popular procedure in esthetic medicine and a variety of other clinical indications [81, 82] with special attention focused on ADSC. These cells can be isolated from subcutaneous fat removed during liposuction procedures after the stromal vascular fraction (SVF) is either mechanically sorted or enzymatically digested with bacterial collagenase, decanted (or centrifuged), washed, and grafted [83]. SVF is composed of cellular components (pre-adipocytes, adipocytes, histiocytes, endothelial cell progenitor cells and ADSCs) and is a rich source of growth factors (i.e., bFGF, IGF-1, VEGFs, PDGF-BB), matrikines, and other paracrine cellular factors. The ADSC secretome has been well characterized, consisting of soluble protein factors and lipid membrane particles (exosomes and ectosomes) that are used internationally in multiple therapeutic clinical trials for a vast array of indications, including dermatologic conditions (esthetics, wound healing, fibrotic diseases, dermatoporosis, etc). It is the loss of ADSC stemness, decreased proliferative potential, and dysfunctional secretome expression accompanying skin aging that continues to draw intense interest [82].

3. Genetic influences on cutaneous cellular senescence

The two major molecular pathways resulting in RS and PS have been observed in the skin [23]. These are reviewed in the following section by examining first the genetic aspects of cutaneous cellular senescence, followed by epigenetic influences. It should be noted here that acquisition of these cellular senescence phenotypes plays a critical role in both normal organismal and tissue level physiology by, for example, dampening fibrotic responses during the remodeling phase of wound repair or suppressing tumor formation [84]. However, it also appears to be a major pathologic driver in age-related disease states [85].

3.1 DNA damage related to telomere biology

The senescent phenotype can be activated by DNA damage at the ends of all eukaryotic chromosomes, called telomeres, which consist of DNA loops containing noncoding repeats of guanine-rich sequences complexed with protective oligomeric proteins (Shelterins). Discovery that chromosomal replicative machinery responsible for somatic cell division cannot synthesize exact duplicates of these structures led to the concept of the 'end-replication problem' during serial passaging [20]; thus, telomeric DNA is subjected to attrition because DNA polymerase fails to replicate the 3′ lagging strands.

Telomeric DNA are shortened by approximately 50–200 bp per cell division and thus a molecular clock is achieved, reflecting the replicative history of primary cells [86]. A specialized DNA polymerase (telomerase) is responsible for fixing the 'end replication problem', maintaining telomeric length, but its expression and function are restricted to immortal postnatal cells; *in vivo,* comprising stem, progenitor, and cancer cells. When cells reach their 'Hayflick Limit' telomeres lose enough DNA [87] to trigger a genomic instability signal and chromosomes become 'uncapped' by loss of Shelterin. This genomic instability signal is a specialized DNA damage response (DDR) and generates telomere dysfunction-induced foci (TIFs). Approximately half of all persistent DNA damage foci are localized to telomeres, and these can trigger RS. But senescent cells can harbor many other forms of persistent chromosomal DNA damage foci, called DNA-SCARS (DNA segments with chromatin alterations reinforcing senescence) [9]. These dynamic structures can also trigger cell cycle arrest and SASP induction.

Independent of telomere length or uncapping by loss of Shelterins, guanine-rich telomeric DNA repeats can become damaged by ROS, generating DDR telomere-associated foci (TAFs), which are associated with triggering the senescence phenotype [19]. This observation has particular relevance to the state of chronic inflammation, SASP expression, and tissue aging (Inflammaging) in skin and other tissues [15, 88, 89], as discussed in Section 3.2.

While epidermal, dermal, and subcutaneous cellular compartments all harbor evidence of RS in aged skin tissue, direct evidence that telomeric DNA associated RS is involved in skin aging is supported by experiments involving ectopic expression of human telomerase (*hTERT*). We reported that neonatal human dermal microvascular endothelial cells (HDMEC) undergo RS *in vitro* but can become immortalized with viral transfer of the catalytic subunit of *hTERT* [90]. Furthermore, these telomerized HDMEC formed fully functional microvessels *in vivo* (perfused with murine blood) that exhibited superior durability with time after xenografting in immunodeficient mice *versus* vessels created with *in vitro*-aged primary HDMEC [91]. As previously reviewed, cutaneous microvasculature of aged papillary dermis is markedly reduced and abnormally structured versus young dermis [38], presenting clinically as telangiectasia and senile purpura/dermatoporosis. The roles of RS, OIS, and other senescent pathways on skin vasculature remain to be determined.

3.2 Genotoxic and exposome insults

As noted in the introduction, cellular senescence can be induced in the absence of any telomeric damage or loss and this premature senescence (PS) has similar deleterious effects on aged tissues, including the skin. The triggers for the PS program generally fall into (a) accumulation of subcytotoxic, unrepairable, non-telomeric DNA damage, including mitochondrial DNA (mtDNA), (b) macromolecular insults to cytosolic and secreted proteins and lipids, and (c) metabolic dysfunction involving an altered mitochondrial-lysosomal axis [92]. All of these PS triggers have been demonstrated in skin cells *in vitro* and *in vivo* [93].

The molecular and cellular effects of chronic UV light exposure (photoaging) have also been well-documented and, in many ways, more extensively than intrinsically aged human skin. Both UVA (320–400 nm, less energy) and UVB (280–320 nm, more energy) light cause photoaging but UVB is mostly absorbed by the epidermis, where it causes sunburns. UVA penetrates the superficial and reticular dermal compartments and is considered a major factor in photoaging. While both UVA and UVB wavelengths generate reactive oxygen species (ROS), indirectly damaging DNA, UVB is also directly mutagenic, causing DNA defects called cyclobutene pyrimidine dimers and 6–4 photoproducts [94]. Remodeling of dermal ECM favoring an atrophic phenotype is triggered in unwounded skin by UV exposure via the activation of mitogen-activated protein kinase (*MAPK*) and activator protein 1 (*AP-1*) signaling pathways which causes downstream expression of matrix metalloproteinases (MMPs) in both the epidermal and dermal compartments [95]. These same pathways block transforming growth factor beta (TGF-β)/SMAD signaling via TGF-b type II receptor down regulation causing decreased collagen synthesis [96, 97]. Dissection of the molecular effects of chronic UV exposure on the DEJ and PD have been recently reviewed [33].

Our understanding of the role senescent cells play in cutaneous aging pathologies continues to evolve. In the past, senescent cells observed in the skin with biomarkers *in vivo* were believed to be passive, unresponsive bystanders recognized morphologically by their enlarged, seemingly flattened, abnormal shapes and senescence-associated (SA) β-galactosidase staining. But characterization of SASP expression in senescent cells (and their paracrine effects) provided compelling evidence that senescent cells are anything but passive.

It is now widely accepted that senescent cells remarkably influence surrounding non-senescent neighbors and ECM networks via secretion of inflammatory cytokines, chemokines, matrikines, MMPs, tissue inhibitors of metalloproteinases (TIMPs), and other proteinase-inhibitor systems that comprise the tissue 'proteinase web' [98]. One such example is the role played by plasminogen activator inhibitor-1 (*PAI-1*) in modulating senescence. *PAI-1* is a soluble and matrix bound serine protease inhibitor with multiple matricellular functions and can be found at increased levels in both dermal fibroblasts from aged donors and premature aging syndrome patients [99–101]. Ectopic expression of *PAI-1* in fibroblasts induces the senescent phenotype and is both necessary and sufficient for RS downstream of p53 [102]. Many other examples of SA ECM alterations have been reviewed [33].

The quintessential example of extrinsic aging involves the postmitotic dermal fibroblast population which responds to 'exposomal' damage [103] by activating DDR pathways, triggering PS and subsequent expression of macromolecular damage profiles involving mtDNA damage. One mechanism of mtDNA damage appears to involve UV light-induced deletion of a significant length of mtDNA, termed the 'common deletion' (CD) [104]. This 49 kb mtDNA fragment contains codons for electron transport chain (ETC) protein complexes I, IV and V which together express 72 ETC subunits, the loss of which cripples physiologic functions of mitochondrial energy metabolism, ROS protective mechanisms, and calcium homeostasis. Tracking the mtDNA CD in human skin revealed that both intrinsic (photo-protected) and extrinsic (chronic UV-damaged) skin contain this marker [105], and that dermal fibroblasts appear to be the culprit for subsequent age-related tissue damage of ECM [106, 107].

3.3 Genetic skin diseases associated with DNA repair pathway defects

Analysis of progeroid syndromes have provided insights into molecular mechanisms of intrinsic and extrinsic skin aging. Common skin phenotypic signs and symptoms shared by both these premature aging disorders and skin aging in the general population include skin atrophy, alopecia, fibrosis, telangiectasia, poikiloderma, canities and both NMSC and melanoma. Rare autosomal recessive patterns of different mutations in DNA repair genes group these heritable disorders into those involving; (1) multiple defects in nucleotide excision repair (NER) genes [e.g., DNA polymerase eta (*POLH*) among six others] coding for repair proteins in xeroderma pigmentosum (XP) [108], (2) transcription and transcription-coupled NER genes in Cockayne syndrome (CS) and (3) mutations in the gene family of RecQ helicases involved in DNA double strand break repair in Werner syndrome, Bloom syndrome, and Rothman-Thomson syndrome [109]. In the latter three disorders, mitochondrial defects have been well documented and correlate with cellular senescence phenotypes [110, 111]. In XP-V null mouse models, loss of *POLH* leads to obesity and marked adipose tissue senescent phenotype expression [112].

3.4 SNPs and transcriptomics

Several genome-wide association studies (GWAS) and meta-analyses performed on young and old populations have identified single nucleotide polymorphisms (SNPs) in genes thought to be correlated with skin aging [113–118] or 'perceived' facial age' [119]. These large cohort-based studies suggest specific allelic variants of pigmentation gene (*MC1R*), aryl hydrocarbon receptor gene (*AHR*), basonuclin 2 gene (*BCN2*), type-1 collagen alpha-2 gene (*Col1A2*) or SNPs within or near the DIAPH2, KCND2 and EDEM1 loci all appear to correlate with both intrinsic and extrinsic skin aging phenotypes and/or youthful skin appearance.

Of all these identified genes and their allelic variants, the biology of *MC1R* gene has received perhaps the most recent attention due to its central role in modulating human (and murine) skin pigmentation systems, the clinical influence of which led to the categorization of Fitzpatrick Skin Phototypes. *MC1R* signaling is associated with both skin cancer and skin aging via its mixed role in UV induced PS in melanocytes and promotion of efficient DNA damage repair [120]. Genetic variants of *MC1R* (coding for G protein-coupled transmembrane melanocortin receptor-1 on melanocytes) are strongly linked to increased risk of both NMSC and melanoma in both red and brown Caucasian phototype cohorts [121]. Meta-analysis of several GWAS studies demonstrated SNPs in or near *MC1R* (and SLC45A2 and IRF4) correlated with different skin aging phenotypes using a skin surface topographic scoring system of solar elastosis [116] and the *MC1R* gene may also affect inflammaging via generation of ROS independent of its function in melanin production [122].

Gene expression studies of the skin aging phenotype have revealed several important observations about the complexities of distinguishing intrinsic from extrinsic mechanisms, as they appear to overlap in many important ways. In human skin, gene profiling and transcriptomic analyses [115, 123–128] have identified thousands of upregulated and downregulated genes in old vs. young and intrinsically aged vs. extrinsically aged (photoaged) skin. One transcriptomic study showed genes associated with mitigating oxidative stress, control of lipid synthesis, and epidermal differentiation were all downregulated in both exposed and photo-protected skin, whereas, elastin expression was increased in exposed skin (consistent with formation of solar elastosis), and interstitial collagen expression decreased in sun protected skin (consistent with intrinsic aging) [127]. Similarly, confirming histologic studies, expression profiling of human aging that spanned subjects between the ages of 24–70 years demonstrated younger-appearing skin upregulated expression of the *LAMA5* gene (DEJ component) and epidermal cell-cell adhesion complex (desmosomal) genes *DSC3* and *CDH1* [126].

Race, sex, and skin tone of subjects also all play a role in the genetic correlates of skin aging. The expression of some aging related genes was found to be sex-dependent in a Caucasian sample [129], and studies in a Han Chinese sample showed distinct genetic variants and phenotypes from that in a Caucasian population [118]. These discrete expression patterns further highlight the complexity of cataloging aging mechanisms in the skin and suggest that much more information would be required from a wider diversity of samples to understand any potential global age-related changes.

Altogether, genetic factors, including telomere DNA loss, genotoxic accumulation of mutations in both genomic and mtDNA, DDR signaling, DNA repair dysfunction, and allelic variations in key cutaneous protective genes controlling pigmentation, inflammation, dermal, epidermal, and subcutaneous physiology all converge on our emerging understanding of the central role cellular senescence plays in skin aging. What follows is a review of how epigenetic factors also influence cellular senescence in cutaneous biology and the aging phenotype.

4. Epigenetic influences on cutaneous cellular senescence

Though the evolution to senescence is usually characterized as a genomically driven phenotype, its manifestation can be largely characterized as an epigenetically entrenched state. The baseline definition of a senescent cell is an otherwise mitotic cell that has entered permanent cell cycle arrest, but this also begets a broader shift in cell behavior and protein production. For these changes to be permanent they must be encoded in long-term gene expression tendencies, i.e., in the cellular epigenome. The epigenome is the composite architecture consisting of chemical and physical modifications to the DNA that do not alter the underlying coding and noncoding sequences but instead modify its oligomeric structure and transcription. These modifications span multiple layers from the local control of specific gene promoters to large scale regulation of entire domains of genes.

Canonically, the epigenome is most strongly associated with cell identity, as it makes accessible the portions of genomic DNA needed for the cell's functional role while segregating and silencing irrelevant regions. Thus, the most dramatic epigenetic shifts are observed when cells differentiate from stem or progenitor states. In this full and dramatic state transition, the function of the cell is redefined, affecting everything from its morphology to its protein production and factor secretion [130]. For skin this can mean, for instance, a transiently amplifying cell in the basal epidermis fully differentiating to a corneocyte through natural turnover or endothelial progenitor cells differentiating into new endothelium in response to an angiogenic signal. Conversely, cells undergo constant but more minor epigenetic events as they are exposed to regular stimuli from the environment, which can lead to upregulation or downregulation of certain behaviors [131]. For instance, methylation sequencing of the same cell type across patients (e.g., epidermal keratinocytes) will show a distribution with perturbation on the mean population value based on internal and local external stimuli [132]. This heterogeneity evolves with different stressors, and with aging itself becomes more prominent. Eventually the stressors can lead to enough diversity to characterize pseudostates and pseudostate transitions that may by-and-large retain the cell's identity but with a different grade of functionality across multiple genes. We will discuss a few additional examples of this in the following sections, like fibrotic versions of connective tissue cells and pro-/anti-inflammatory versions of macrophages; however, the focus of this section that could also be considered an epigenetic pseudostate is senescence. It meets the criteria in that it still retains core cell identity but also dramatically affects a multitude of genes to alter protein production and secretion profile, and thus requires a core epigenetic component. In this section we will review the multitude of epigenetic changes that accompanies this pseudostate evolution in skin cells, what role they play in establishing the senescent phenotype and how they may potentially be engaged therapeutically.

4.1 Sequence specific modulation

One modality of epigenetic is sequence specific meaning it targets specific regions of the genetic code—either DNA or RNA. DNA base pair methylation is a well-known example of this modality. Cytosine is the most commonly methylated base in eukaryotic cells and when methylated, often serves to block the activity of RNA polymerase, as in the context of CpG islands. Found in the promoter region of many genes, CpG islands are clusters of methylated cytosine followed by guanine, wherein the methylation inhibits (silences) the transcription of that gene [132]. Widespread hypomethylation has been documented in aging and senescent fibroblasts, and in some cases, impairs cell cycling pathways through the suppression of cyclin pathways. Specifically, a lack of methylated sites leads to the upregulation of *p16INK4a* which inhibits cyclin D/CDK4 to suppress G1 phase progression, while upregulation of *p14ARF* leads to activation of p53/p21 and inhibits cyclin E/CDK4 to prevent S phase progression [7]. The global methylation status of fibroblasts is directly and strongly correlated with donor chronological age through regression algorithms known as 'epigenetic clocks.' These algorithms calculate a weighted linear combination of the beta coefficients (the percent signal from the methylated out of the total unmethylated and methylated alleles) [133]. When dermal fibroblasts were passaged towards replicative senescence (RS) these epigenetic clocks show aggregated methylomic evolution. The cell cycle was reengaged by overexpressing the telomerase gene *hTERT*, causing cells to progress to further doubling. However, the epigenetic clock did not reverse and the cells continued to age, bypassing RS, further hinting that a broader epigenetic change was occurring through the progression to senescence rather than just the suppression of a few mitotic arrest genes [134].

Conversely, sequence specific epigenetic regulation on the level of transcribed RNA is accomplished through feedback mechanisms by families of non-coding RNA-including microRNAs, siRNAs, long and short non-coding RNAs, and others. These non-coding RNAs will interact with other DNA, RNA, and proteins to regulate their expression, further enhancing the complexity of the transcriptome over the more rigid landscape of the methylome [135]. Thus, non-coding species are often used to not just reinforce but also propagate the senescence response. The particular influence of miRNAs, short sequences that complement and bind to specific regions of mRNAs to limit their stability and thus their translation likelihood, has been explored in the context of cutaneous cell senescence [136]. For instance, UV-induced senescent fibroblasts are known to produce miR-34 which targets a number of transcripts within these cells for cell cycle regulatory genes like *MYC* and *BCL2* as well as genes for other epigenetic factors such as E2H and SIRT1 [137].

Meanwhile, in wounding-induced senescence the extracellular secretion of miR-21 as part of the SASP phenotype triggers the activation of resident macrophages to drive the local inflammatory response [138], but these represent only a few of a handful of drivers. Senescent keratinocytes, for instance, have displayed upregulation of over a hundred different microRNAs correlated with expression of the senescence biomarkers p16, p53, and senescence-associated β-galactosidase (SA-β-Gal) [136]. Together, these mechanisms represent the precise regulation of specific genomic targets and interfering with transcription machinery as one mode of enforcing the senescent epigenetic state.

4.2 Compaction

For regulation across gene domains (~150 base pairs or greater), the epigenome uses methods of physical compaction to close off regions of the genome from transcription. The negative charge of the DNA attracts it to wrap around the positively charged protein octamer spools called histones, which segregate the sequences away from transcription machinery. Chemical modifications like methylation, acetylation, and ubiquitination of the amino acid residues on the tails of these histone proteins alter the charge interaction with DNA and with other histones influencing oligomeric structure [139]. Senescence engages in this mechanism by modulating the enzymatic activity that regulates these histone tails. For example, the activity of methyltransferases like EZH2, which adds trimethylation to the lysine residue 27 of histone 3 (H2K27me3), is reduced in senescent cells. This reduction in the resulting H3K27me3, especially at the *INK4a/ARF* locus mentioned previously, reinforces the discontinuation of the cell cycle [8]. Other forms of histone tail modification include acetylation, which tends to promote more transcription. One of the most well-studied classes of deacetylation enzymes is the sirtuin family of proteins. In both fibroblasts and keratinocytes, Sirt1 and Sirt6 directly respond to DNA damage and inflammation, but their expression is diminished in senescent cells [140]. Interestingly, both Sirt1 and Sirt6 also play an active role in regulating collagen balance, thus their downregulation could be conceptually likened to senescence of the dermal ECM and its turnover, just like that of cellular turnover.

Histones can also be modified through changes within the core octamer proteins themselves and a hallmark example of this phenomenon is the variant species of the H2A protein known as H2A.J. This modified protein is prevalent in a lot of senescent skin cell types where it weakens the binding of another histone in the complex, H1, triggering a signaling cascade that preempts the interferon response and contributes to initiation of SASP expression [141]. In senescent epidermal keratinocytes in particular, the increase in H2A.J variants is correlated with arrested cell cycle and maturation of the basal cells into mature corneocytes, thus it may play a direct role in the morphologic phenomena of epidermal thinning seen with age [142]. The broader contribution of these histone changes, along with local DNA methylation shifts, is the transition to wide-reaching genome compaction in senescent cells, for example the condensation of senescence associated heterochromatin foci, as in H3K9me3 rich regions of nuclease resistant compact facultative heterochromatin [11]. These foci are seen across skin cell types like fibroblasts and keratinocytes and are thought to entrench the senescent state by long term segregation and silencing of mitotic genes [143]. However, the evolution of these foci seems to be specific to the type of senescence induction, most prominent in OIS, suggesting that senescence itself may even be a family of pseudo-states rather than a distinct, singular manifestation [144]. Nevertheless, in general, these forms of epigenetic modification which bias entire regions of genes from active to passive and vice versa truly embody a cell state/pseudostate.

4.3 Alternative epigenetic pseudostates

The natural and prevalent engagement of senescence, even in young tissues, reflects its role as a form of stress response. In fact, a major function of senescence is to prevent the evolution of alternate, more detrimental states of the cells and tissue under these conditions. One such competing epigenetic pseudostate is fibrosis. The fibrotic transition is a common feature in the pathological evolution of many tissues, i.e., hypertrophic scarring and keloids in the skin, idiopathic pulmonary fibrosis in the lung, cirrhosis in the liver [145]. A key component of fibrosis is the differentiation of various cell types including fibroblasts, adipocytes, epithelial cells, and endothelial cells into a population known as myofibroblasts [146]. As mentioned, differentiation is canonically an epigenetic event as cells convert and specify their functional gene regions while silencing other unused regions. It involves the same modalities of control—methylation, histone tags, chromatin structure, etc.—often with more dramatic and permanent modifications. These activated myofibroblasts are critical for the repair response in that they secrete superfluous extracellular matrix (ECM) components (Collagen 1, alpha-smooth muscle actin (α-SMA), fibronectin, etc) that accumulate in the connective tissue [147]. At the same time, these cells diminish the process of anabolic degradation of ECM through reduction of MMPs [148]. Unbridled overgrowth of these myofibroblasts, as evidenced by the overactivation of growth factors like connective tissue growth factor (CTGF), leads to the buildup and disorganization of the connective tissue [149]. Senescence in this context is thought to be a responsive, secondary epigenetic evolution that is engaged to shut down this population and stop the overgrowth [150]. These processes—from the epigenetic cell identity shift (e.g., epithelial-to-mesenchymal or fibroblast-to-myofibroblast transitions, depending on the starting cell types) to the epigenetic proliferation-suppressed state (induction of senescence)—represent relatively fast epigenetic turnovers. As such, a key mediator of this rapid transition is thought to be the slew of non-coding RNAs, like let-7 g to engage TGFbeta driven myoblast conversion and miR-127-3p to induce p53/p21 drivers of senescence [151, 152].

Another alternative pseudostate that competes with senescence is of course cancer and more particularly for skin, melanoma. Like senescence, cancer is a state transition that involves bypassing apoptotic pathways, yet these aberrant cells also bypass the suppression of their cell cycle gene networks [14]. It is thought that melanoma cells are able to undo the senescence epigenetics and re-engage the cell cycle due to the deleterious recruitment of epigenetic enzymes, like histone demethylases and Jumonji proteins [153]. This means a host of pathways whose methylation would otherwise lead to cell cycle suppression, like the p15INK4B or the p27Kip1 pathways, are methylated without cell cycle arrest in melanoma [154–156]. The use of inhibitors to target these epigenetic enzymes seems to be a promising methodology to restore the cell cycle arrest and control the cancerous growth [157].

An interesting intersection of epigenetic and oncogenic pseudostates is highlighted in Merkel cell carcinoma (MCC). This aggressive, non-melanoma skin cancer is rare but occurs primarily in the elderly and immunosuppressed. Interestingly, methylation clock analysis of MCC cells shows their epigenetic age as significantly younger than the chronologic age of the patients from which they were derived—a stark contrast from the continually progressing epigenetic age of senescent cells. Further analysis of these MCC cells did not indicate any signs of pluripotency [64]. The mechanism by which MCCs reverse their epigenetic age is still unknown, however, it may be related to other epigenetic alterations recently discovered in this cell type, including decreased H3K27me3 expression [158, 159] and overactivity of the lysine-specific histone demethylase 1A [160, 161]. These are just some examples of this fundamental need to tightly control and disengage mitotic networks and why senescence requires a complex regulatory architecture like the epigenome.

4.4 Enablers of senescence

The phenomenon of senescence is promoted by the epigenetics of not just the arrested cells in question, but also that of the other resident cells that enable this transition. Though senescence is thought to be a permanent state, the persistence of senescence in the tissue is only meant to be transient. This is because the key function of this state is to respond to stressors by retaining cells, despite their damage, to maintain the tissue temporarily while preventing them going down the more detrimental alternate routes mentioned, all the while signaling the immune system and other repair mechanisms. When the immune system is young and efficient its cells are recruited to the skin and other tissues to clear out the senescent cells [162]. With aging, however, the number and lifetime of these senescent populations increases due to the altered epigenetic pseudostates of the senescent clearing cells as well, contributing to innate immunosurveillance dysfunction of the skin. One example of this is in the dominance of the pro-inflammatory M1 macrophage pseudostate over the anti-inflammatory M2 macrophage pseudostate [163]. There are a number of histone methylation and acetylation modifiers that play a role in pseudostate fate decision, for instance histone deacetylase 3 promoting M1 macrophages or the SYMD family of methyltransferases promoting M2 macrophages [164]. With the accumulation of stressors over a lifetime, the more pro-inflammatory epigenetic pseudo-states are favored in skin and other tissues, especially in response to factors like SASP or inflammaging [165]. In addition, in some disease states like type 2 diabetes, the wound healing response and inflammation tends to exaggerate the M1 state response with focal DNA methylation components at sites like peroxisome proliferator activated receptor gamma (PPARγ) or and elevation of miR-125b [164]. This epigenetic shift in the balance of macrophage cells then ties back to senescence as the M1 macrophage predominantly engages in more phagocytic clearance of foreign pathogens, while the M2 macrophages carry out more phagocytic clearance of damaged host cells (efferocytosis) [166]. This, coupled with the fact that senescent cells develop ways to better evade apoptosis, means that they are more likely to accumulate [167] in aged tissue. There are additional immune cell types that are similarly driven by the pro-inflammatory transition, yet become impaired at senescent cell clearance, including NK cells and neutrophils [167]. Altogether, this epigenetic evolution of the regulator cells, part of inflammaging, proves just as critical to the manifestation of a sustained senescence pressure in cutaneous tissue as epigenetic changes engaged in the non-dividing cells themselves.

4.4.1 Distinction from temporary cell cycle arrest

Though sometimes associated with senescence, somatic stem cells (as opposed to differentiated cells) are typically associated with another form of cell cycle arrest, known as quiescence. Because their role is to remain as a niched tissue reserve, they often enter periods of temporary cell cycle arrest with a prolonged G0, instead of a permanent one, until they are called to activate, to proliferate and differentiate, by a stressor [168]. One major epigenetic distinction that enables this temporary quiescence vs. permanent senescence is the utilization of bivalent domains. These are regions of genes that are regulated by both a repressive histone tag as well as an activating one, that allows the region to rapidly switch from one state to another depending on stimulus [169]. A prime example of the use of this is in the coinciding utilization of repressive H3K27me3 and activating H3K4me3, which maintains a tenuous baseline suppression of the gene region. This pair forming a bivalent domain is widely used throughout the embryonic stem cell genome, establishing its broad potency as a cell type with the potential to express a lot of different proteins [170]. But when the same domains were searched for in dermal hair follicle stem cells (HFSC), they were found to be substantially restricted to lineage-specific factors like *Sox9* and *Nfatc1* and growth factor *FGF18* [171]. Then, when these HFSC were stimulated to activate, many of the genes with H3K4me3 activating markers, which are located primarily near the gene promoters, were further reinforced by additional H3K79 dimethylation in the gene body, to tip the scale from suppression to activation [171]. These genes included many cell cycle regulators which, when combined with the cell lineage factors, properly executed differentiation. Thus, this mechanism of readily switchable suppression to expression establishes a major distinction in epigenetic regulation from cell cycle in quiescence from that of senescence where the cell cycle genes are more permanently, epigenetically suppressed.

4.4.2 Manipulability of senescent epigenetics

Earlier, we mentioned how drugs targeting epigenetic enzymes represent one methodology for modulating some of the epigenetic changes that drive senescence, such as senolytic therapies. However, a broader and more dramatic approach of epigenetic evolution is through the process of cellular reprogramming. This technology was inspired by the core epigenetic reset that occurs during the process of reproduction in which sperm and egg, two cells with very precise roles and epigenetic identities, are reprogrammed to make embryonic cells—epigenetically plastic cells that can differentiate into any cell in the body. The isolation and recapitulation of this process in any desired cell type was achieved through the discovery of core transcription factors [172]. When overexpressed in cells, this set of core transcription factors would drive a full epigenetic remodeling to produce embryonic-like cells with all their differentiation potential. This process is called induced pluripotent stem cell reprogramming (iPSC) and has been utilized in a variety of different cell types with dermal fibroblasts being the gold standard for many studies [173]. Even fully senescent fibroblast populations established from 51 population doublings and maintained for two months in culture, successfully showed iPSC reprogramming, as evidenced by revived proliferation, reduced p16 and p21, and re-differentiation after reaching the pluripotent state [174]. Crucially, the re-differentiated progeny were once again able to be passaged into senescence, thus suggesting that malignant transformation was not induced during the entire process. Furthermore, one of the key reprogramming factors *Oct4*, has been shown to independently re-engage senescent hair follicle mesenchymal stem cells back into cycling by engaging a host of DNA methyltransferase to inhibit the p21 pathway [175]. More recently researchers have shown that the prevalence of senescence in a population can be reduced with even a transient application of the reprogramming factors [176–178]. Though whether this means a re-engagement of senescent cells in the cell cycle or simply competitive growth advantage of healthy cells remains to be seen. This represents an enticing new possibility in that epigenetic manipulation may possibly counter the accumulation of senescent cells in many aged and diseased tissues, including the skin.

5. Conclusion

The skin represents an excellent organ system in which the effects of cellular senescence manifest as observed clinical changes in organismal health and disease.

A myriad of processes drives the genomic erosion that instigates the transition to senescence. Some of these processes are more stereotyped, engineered into the cell by design, and are observed in chronologically aged skin, while others are stochastic and driven by environmental conditions, exemplified by exposomal damage. Either way, the result is an evolution of the entire state of the cell. This means more than just the direct arrestation of the cell cycle, but also entails changes through the cellular transcriptome, proteome, and secretome as encoded by alterations to the core cellular epigenome. This also involves a myriad of changes to the many layers of architecture that encode a cell's function and identity. The global process is critical for the skin's ability to retain functional integrity upon stress and insult as the first line of defense for the body, and in many ways, senescence represents the least of multiple evils.

This review gives a glimpse of how and why intrinsic and extrinsic factors trigger cutaneous cellular senescent phenotypes, leaving several important questions unanswered. For example, which genetic and epigenetic factors determine the dominant decision pathways favoring senescence vs. apoptosis or any other disease states for different skin compartments? How are the various types of senescence manifestations comparable in terms of evolution and manipulability? What are the molecular and cellular consequences of therapeutic re-engagement of senescent cells into the cell cycle? As the focus on aging grows as an ever more prominent factor in clinical and investigative dermatology, insights on these questions into the nature of senescence become a critical step towards both dermatologic therapeutic advancement specifically and translational medicine in general.

Acknowledgements

Intellectual and financial support by Turn Biotechnologies, Inc. is gratefully appreciated.

Conflict of interest

The authors declare no conflict of interest.

Author details

Tapash Jay Sarkar, Maiko Hermsmeier, Jessica L. Ross and G. Scott Herron*
Turn Biotechnologies, Inc., Mountain View, California, USA

*Address all correspondence to: scott.herron@turn.bio

References

[1] Rodier F, Campisi J. Four faces of cellular senescence. The Journal of Cell Biology. 2011;**192**(4):547-556

[2] Gorgoulis V, Adams PD, Alimonti A, Bennett DC, Bischof O, Bishop C, et al. Cellular senescence: Defining a path forward. Cell. 2019;**179**(4):813-827

[3] Hayflick L, Moorhead PS. The serial cultivation of human diploid cell strains. Experimental Cell Research. 1961;**25**(3):585-621

[4] Kuilman T, Michaloglou C, Mooi WJ, Peeper DS. The essence of senescence. Genes & Development. 2010;**24**(22): 2463-2479

[5] Coppé J-P, Patil CK, Rodier F, Sun Y, Muñoz DP, Goldstein J, et al. Senescence-associated secretory phenotypes reveal cell-nonautonomous functions of oncogenic RAS and the p53 tumor suppressor. PLoS Biology. 2008;**6**(12): 2853-2868

[6] Coppé J-P, Patil CK, Rodier F, Krtolica A, Beauséjour CM, Parrinello S, et al. A human-like senescence-associated secretory phenotype is conserved in mouse cells dependent on physiological oxygen. PLoS One. 2010;**5**(2):e9188

[7] Li X, Xu H, Xu C, Lin M, Song X, Yi F, et al. The Yin-Yang of DNA damage response: Roles in tumorigenesis and cellular senescence. International Journal of Molecular Sciences. 2013;**14**(2): 2431-2448

[8] Criscione SW, De Cecco M, Siranosian B, Zhang Y, Kreiling JA, Sedivy JM, et al. Reorganization of chromosome architecture in replicative cellular senescence. Science Advances. 2016;**2**(2):e1500882

[9] Rodier F, Muñoz DP, Teachenor R, Chu V, Le O, Bhaumik D, et al. DNA-SCARS: Distinct nuclear structures that sustain damage-induced senescence growth arrest and inflammatory cytokine secretion. Journal of Cell Science. 2011;**124**(1):68-81

[10] Prieto LI, Graves SI, Baker DJ. Insights from in vivo studies of cellular senescence. Cell. 2020;**9**(4):954

[11] Corpet A, Stucki M. Chromatin maintenance and dynamics in senescence: A spotlight on SAHF formation and the epigenome of senescent cells. Chromosoma. 2014;**123**(5):423-436

[12] Lundberg AS, Hahn WC, Gupta P, Weinberg RA. Genes involved in senescence and immortalization. Current Opinion in Cell Biology. 2000;**12**(6): 705-709

[13] Wang E. Senescent human fibroblasts resist programmed cell death, and failure to suppress bcl2 is involved. Cancer Research. 1995;**55**(11):2284-2292

[14] Childs BG, Baker DJ, Kirkland JL, Campisi J, van Deursen JM. Senescence and apoptosis: Dueling or complementary cell fates? EMBO Reports. 2014;**15**(11):1139-1153

[15] Salvioli S, Monti D, Lanzarini C, Conte M, Pirazzini C, Bacalini MG, et al. Immune system, cell senescence, aging and longevity—Inflamm-aging reappraised. Current Pharmaceutical Design. 2013;**19**(9):1675-1679

[16] Xu W, Wong G, Hwang YY, Larbi A. The untwining of immunosenescence and aging. Seminars in Immunopathology. 2020;**42**(5):559-572

[17] Hayflick L. The limited in vitro lifetime of human diploid cell strains.

Experimental Cell Research. 1965;**37**: 614-636

[18] Bodnar AG, Ouellette M, Frolkis M, Holt SE, Chiu CP, Morin GB, et al. Extension of life-span by introduction of telomerase into normal human cells. Science. 1998;**279**(5349):349-352

[19] d'Adda di Fagagna F, Reaper PM, Clay-Farrace L, Fiegler H, Carr P, Von Zglinicki T, et al. A DNA damage checkpoint response in telomere-initiated senescence. Nature. 2003;**426**(6963):194-198

[20] Shay JW, Wright WE. Telomeres and telomerase: Three decades of progress. Nature Reviews. Genetics. 2019;**20**(5):299-309

[21] Serrano M, Lin AW, McCurrach ME, Beach D, Lowe SW. Oncogenic ras provokes premature cell senescence associated with accumulation of p53 and p16INK4a. Cell. 1997;**88**(5):593-602

[22] Wang AS, Dreesen O. Biomarkers of cellular senescence and skin aging. Frontiers in Genetics. 2018;**9**:247

[23] Ho CY, Dreesen O. Faces of cellular senescence in skin aging. Mechanisms of Ageing and Development. 2021;**198**:111525

[24] Farage MA, Miller KW, Elsner P, Maibach HI. Characteristics of the aging skin. Advances in Wound Care. 2013;**2**(1):5-10

[25] Wiegand C, Raschke C, Elsner P. Skin aging: A brief summary of characteristic changes. In: Farage MA, Miller KW, Maibach HI, editors. Textbook of Aging Skin [Internet]. Berlin, Heidelberg: Springer; 2017. pp. 55-65. DOI: 10.1007/978-3-662-47398-6_5

[26] Montagna W. Comparative anatomy and physiology of the skin. Archives of Dermatology. 1967;**96**(4):357-363

[27] Blanpain C, Fuchs E. Epidermal stem cells of the skin. Annual Review of Cell and Developmental Biology. 2006;**22**:339-373

[28] Russell-Goldman E, Murphy GF. The pathobiology of skin aging: New insights into an old dilemma. The American Journal of Pathology. 2020;**190**(7):1356-1369

[29] Branchet MC, Boisnic S, Frances C, Robert AM. Skin thickness changes in normal aging skin. Gerontology. 1990;**36**(1):28-35

[30] Sauermann K, Clemann S, Jaspers S, Gambichler T, Altmeyer P, Hoffmann K, et al. Age related changes of human skin investigated with histometric measurements by confocal laser scanning microscopy in vivo. Skin Research and Technology. 2002;**8**(1):52-56

[31] Wurm EMT, Longo C, Curchin C, Soyer HP, Prow TW, Pellacani G. In vivo assessment of chronological ageing and photoageing in forearm skin using reflectance confocal microscopy. The British Journal of Dermatology. 2012;**167**(2):270-279

[32] Haytoglu NSK, Gurel MS, Erdemir A, Falay T, Dolgun A, Haytoglu TG. Assessment of skin photoaging with reflectance confocal microscopy. Skin Research and Technology. 2014;**20**(3):363-372

[33] Roig-Rosello E, Rousselle P. The human epidermal basement membrane: A shaped and cell instructive platform that aging slowly alters. Biomolecules. 2020;**10**(12):E1607

[34] Reilly DM, Lozano J. Skin collagen through the lifestages: Importance for skin health and beauty. Plastic and Aesthetic Research [Internet]. 2021;**8**:2. Available from: https://parjournal.net/article/view/3863

[35] Waller JM, Maibach HI. Age and skin structure and function, a quantitative approach (II): Protein, glycosaminoglycan, water, and lipid content and structure. Skin Research and Technology: Official Journal of International Society for Bioengineering and the Skin (ISBS) [and] International Society for Digital Imaging of Skin (ISDIS) [and] International Society for Skin Imaging (ISSI) Journal. 2006;**12**(3):145-154

[36] Braverman IM. Elastic fiber and microvascular abnormalities in aging skin. Clinics in Geriatric Medicine. 1989;**5**(1):69-90

[37] Li L, Mac-Mary S, Sainthillier J-M, Nouveau S, de Lacharriere O, Humbert P. Age-related changes of the cutaneous microcirculation in vivo. Gerontology. 2006;**52**(3):142-153

[38] Bentov I, Reed MJ. The effect of aging on the cutaneous microvasculature. Microvascular Research. 2015;**100**:25-31

[39] Fenske NA, Lober CW. Structural and functional changes of normal aging skin. Journal of the American Academy of Dermatology. 1986;**15**(4 Pt 1):571-585

[40] Page ME, Lombard P, Ng F, Göttgens B, Jensen KB. The epidermis comprises autonomous compartments maintained by distinct stem cell populations. Cell Stem Cell. 2013;**13**(4): 471-482

[41] Rognoni E, Watt FM. Skin cell heterogeneity in development, wound healing, and cancer. Trends in Cell Biology. 2018;**28**(9):709-722

[42] Eckhart L, Zeeuwen PLJM. The skin barrier: Epidermis vs environment. Experimental Dermatology. 2018;**27**(8):805-806

[43] Liu N, Matsumura H, Kato T, Ichinose S, Takada A, Namiki T, et al. Stem cell competition orchestrates skin homeostasis and ageing. Nature. 2019;**568**(7752):344-350

[44] Tigges J, Krutmann J, Fritsche E, Haendeler J, Schaal H, Fischer JW, et al. The hallmarks of fibroblast ageing. Mechanisms of Ageing and Development. 2014;**138**:26-44

[45] Wlaschek M, Maity P, Makrantonaki E, Scharffetter-Kochanek K. Connective tissue and fibroblast senescence in skin aging. The Journal of Investigative Dermatology. 2021;**141**(4):985-992

[46] Moreiras H, Seabra MC, Barral DC. Melanin transfer in the epidermis: The pursuit of skin pigmentation control mechanisms. International Journal of Molecular Sciences. 2021;**22**(9):4466

[47] Bennett DC, Medrano EE. Molecular regulation of melanocyte senescence. Pigment Cell Research. 2002;**15**(4):242-250

[48] Waaijer MEC, Gunn DA, Adams PD, Pawlikowski JS, Griffiths CEM, van Heemst D, et al. P16INK4a positive cells in human skin are indicative of local elastic fiber morphology, facial wrinkling, and perceived age. Journal of Gerontology Series A. 2016;**71**(8):1022-1028

[49] Victorelli S, Lagnado A, Halim J, Moore W, Talbot D, Barrett K, et al. Senescent human melanocytes drive skin ageing via paracrine telomere dysfunction. The EMBO Journal. 2019;**38**(23):e101982

[50] West HC, Bennett CL. Redefining the role of langerhans cells as immune regulators within the skin. Frontiers in Immunology. 2018;**8**:1941

[51] Nguyen AV, Soulika AM. The dynamics of the skin's immune system. International Journal of Molecular Sciences. 2019;**20**(8):E1811

[52] Ali N, Zirak B, Rodriguez RS, Pauli ML, Truong H-A, Lai K, et al. Regulatory T cells in skin facilitate epithelial stem cell differentiation. Cell. 2017;**169**(6):1119-1129.e11

[53] Pilkington SM, Bulfone-Paus S, Griffiths CEM, Watson REB. Inflammaging and the skin. The Journal of Investigative Dermatology. 2021;**141**(4):1087-1095

[54] Pilkington SM, Dearman RJ, Kimber I, Griffiths CEM. Langerhans cells express human β-defensin 3: Relevance for immunity during skin ageing. The British Journal of Dermatology. 2018;**179**(5):1170-1171

[55] Pereira BI, Devine OP, Vukmanovic-Stejic M, Chambers ES, Subramanian P, Patel N, et al. Senescent cells evade immune clearance via HLA-E-mediated NK and CD8+ T cell inhibition. Nature Communications. 2019;**10**(1):2387

[56] Moro F, Fania L, Sinagra JLM, Salemme A, Di Zenzo G. Bullous pemphigoid: Trigger and predisposing factors. Biomolecules. 2020;**10**(10):1432

[57] Maraee A, Farag AGA, Gadallah MM, Abdou AG. Tumour-infiltrating Langerhans cells in non-melanoma skin cancer, a clinical and immunohistochemical study. Ecancermedicalscience. 2020;**14**:1045

[58] Gerlini G, Di Gennaro P, Mariotti G, Urso C, Chiarugi A, Caporale R, et al. Human langerhans cells are immature in melanoma sentinel lymph nodes. Blood. 2012;**119**(20):4807-4808

[59] Abraham J, Mathew S. Merkel cells: A collective review of current concepts. International Journal of Applied & Basic Medical Research. 2019;**9**(1):9-13

[60] Skedung L, El Rawadi C, Arvidsson M, Farcet C, Luengo GS, Breton L, et al. Mechanisms of tactile sensory deterioration amongst the elderly. Scientific Reports. 2018;**8**(1): 5303

[61] García-Piqueras J, García-Mesa Y, Cárcaba L, Feito J, Torres-Parejo I, Martín-Biedma B, et al. Ageing of the somatosensory system at the periphery: Age-related changes in cutaneous mechanoreceptors. Journal of Anatomy. 2019;**234**(6):839-852

[62] Schadendorf D, Lebbé C, Zur Hausen A, Avril M-F, Hariharan S, Bharmal M, et al. Merkel cell carcinoma: Epidemiology, prognosis, therapy and unmet medical needs. European Journal of Cancer (Oxford, England: 1990). 1990;**2017**(71):53-69

[63] Harms PW, Harms KL, Moore PS, DeCaprio JA, Nghiem P, Wong MKK, et al. The biology and treatment of Merkel cell carcinoma: Current understanding and research priorities. Nature Reviews. Clinical Oncology. 2018;**15**(12):763-776

[64] Chteinberg E, Vogt J, Kolarova J, Bormann F, van den Oord J, Speel EJ, et al. The curious case of Merkel cell carcinoma: Epigenetic youth and lack of pluripotency. Epigenetics. 2020;**15**(12):1319-1324

[65] Halfter W, Oertle P, Monnier CA, Camenzind L, Reyes-Lua M, Hu H, et al. New concepts in basement membrane biology. The FEBS Journal. 2015;**282**(23):4466-4479

[66] Weihermann AC, Lorencini M, Brohem CA, de Carvalho CM. Elastin structure and its involvement in skin photoageing. International Journal of Cosmetic Science. 2017;**39**(3):241-247

[67] Hu MS, Moore AL, Longaker MT. A fibroblast is not a fibroblast is not a fibroblast. The Journal of Investigative Dermatology. 2018;**138**(4):729-730

[68] Philippeos C, Telerman SB, Oulès B, Pisco AO, Shaw TJ, Elgueta R, et al. Spatial and single-cell transcriptional profiling identifies functionally distinct human dermal fibroblast subpopulations. The Journal of Investigative Dermatology. 2018;**138**(4):811-825

[69] Mine S, Fortunel NO, Pageon H, Asselineau D. Aging alters functionally human dermal papillary fibroblasts but not reticular fibroblasts: A new view of skin morphogenesis and aging. PLoS One. 2008;**3**(12):e4066

[70] Yamada T, Hasegawa S, Miyachi K, Date Y, Inoue Y, Yagami A, et al. Laminin-332 regulates differentiation of human interfollicular epidermal stem cells. Mechanisms of Ageing and Development. 2018;**171**:37-46

[71] Griffin MF, des Jardins-Park HE, Mascharak S, Borrelli MR, Longaker MT. Understanding the impact of fibroblast heterogeneity on skin fibrosis. Disease Models & Mechanisms. 2020;**13**(6). DOI: 10.1242/dmm.044164

[72] Mascharak S, des Jardins-Park HE, Davitt MF, Griffin M, Borrelli MR, Moore AL, et al. Preventing Engrailed-1 activation in fibroblasts yields wound regeneration without scarring. Science [Internet]. 2021;**372**(6540):eaba2374. Available from: https://www.science.org/doi/abs/10.1126/science.aba2374

[73] Giralt M, Cereijo R, Villarroya F. Adipokines and the endocrine role of adipose tissues. In: Herzig S, editor. Metabolic Control [Internet], Handbook of Experimental Pharmacology. Cham: Springer International Publishing; 2016. pp. 265-282. DOI: 10.1007/164_2015_6

[74] Cai Y, Li J, Jia C, He Y, Deng C. Therapeutic applications of adipose cell-free derivatives: A review. Stem Cell Research & Therapy. 2020;**11**(1):312

[75] Guo S, Wang T, Zhang S, Chen P, Cao Z, Lian W, et al. Adipose-derived stem cell-conditioned medium protects fibroblasts at different senescent degrees from UVB irradiation damages. Molecular and Cellular Biochemistry. 2020;**463**(1-2):67-78

[76] Tchkonia T, Morbeck DE, Von Zglinicki T, Van Deursen J, Lustgarten J, Scrable H, et al. Fat tissue, aging, and cellular senescence. Aging Cell. 2010;**9**(5):667-684

[77] Palmer AK, Xu M, Zhu Y, Pirtskhalava T, Weivoda MM, Hachfeld CM, et al. Targeting senescent cells alleviates obesity-induced metabolic dysfunction. Aging Cell. 2019;**18**(3):e12950

[78] Liao N, Shi Y, Zhang C, Zheng Y, Wang Y, Zhao B, et al. Antioxidants inhibit cell senescence and preserve stemness of adipose tissue-derived stem cells by reducing ROS generation during long-term in vitro expansion. Stem Cell Research & Therapy. 2019;**10**(1):306

[79] Kim M, Park KY, Lee M-K, Jin T, Seo SJ. Adiponectin suppresses UVB-induced premature senescence and hBD2 overexpression in human keratinocytes. PLoS One. 2016;**11**(8):e0161247

[80] Ezure T, Amano S. Adiponectin and leptin up-regulate extracellular matrix production by dermal fibroblasts. BioFactors (Oxford, England). 2007;**31**(3-4):229-236

[81] Gaur M, Dobke M, Lunyak VV. Mesenchymal stem cells from adipose tissue in clinical applications for dermatological indications and skin aging. International Journal of Molecular Sciences. 2017;**18**(1):208

[82] Trzyna A, Banaś-Ząbczyk A. Adipose-derived stem cells secretome and its potential application in "stem cell-free therapy". Biomolecules. 2021;**11**(6):878

[83] Cleveland EC, Albano NJ, Hazen A. Roll, spin, wash, or filter? Processing of lipoaspirate for autologous fat grafting: An updated, evidence-based review of the literature. Plastic and Reconstructive Surgery. 2015;**136**(4):706-713

[84] Demaria M, Ohtani N, Youssef SA, Rodier F, Toussaint W, Mitchell JR, et al. An essential role for senescent cells in optimal wound healing through secretion of PDGF-AA. Developmental Cell. 2014;**31**(6):722-733

[85] McHugh D, Gil J. Senescence and aging: Causes, consequences, and therapeutic avenues. The Journal of Cell Biology. 2018;**217**(1):65-77

[86] Harley CB, Futcher AB, Greider CW. Telomeres shorten during ageing of human fibroblasts. Nature. 1990;**345**(6274):458-460

[87] Blackburn EH, Challoner PB. Identification of a telomeric DNA sequence in *Trypanosoma brucei*. Cell. 1984;**36**(2):447-457

[88] Herbig U, Ferreira M, Condel L, Carey D, Sedivy JM. Cellular senescence in aging primates. Science. 2006;**311**(5765):1257

[89] Jeyapalan JC, Ferreira M, Sedivy JM, Herbig U. Accumulation of senescent cells in mitotic tissue of aging primates. Mechanisms of Ageing and Development. 2007;**128**(1):36-44

[90] Yang J, Chang E, Cherry AM, Bangs CD, Oei Y, Bodnar A, et al. Human endothelial cell life extension by telomerase expression. The Journal of Biological Chemistry. 1999;**274**(37):26141-26148

[91] Yang J, Nagavarapu U, Relloma K, Sjaastad MD, Moss WC, Passaniti A, et al. Telomerized human microvasculature is functional in vivo. Nature Biotechnology. 2001;**19**(3):219-224

[92] Kim M, Park HJ. Molecular mechanisms of skin aging and rejuvenation [Internet]. In: Molecular Mechanisms of the Aging Process and Rejuvenation. Rejika: IntechOpen; 2016. Available from: https://www.intechopen.com/chapters/50471

[93] Fitsiou E, Pulido T, Campisi J, Alimirah F, Demaria M. Cellular senescence and the senescence-associated secretory phenotype as drivers of skin photoaging. The Journal of Investigative Dermatology. 2021;**141**(4):1119-1126

[94] Schuch AP, Moreno NC, Schuch NJ, Menck CFM, Garcia CCM. Sunlight damage to cellular DNA: Focus on oxidatively generated lesions. Free Radical Biology & Medicine. 2017;**107**:110-124

[95] Quan T, Little E, Quan H, Qin Z, Voorhees JJ, Fisher GJ. Elevated matrix metalloproteinases and collagen fragmentation in photodamaged human skin: Impact of altered extracellular matrix microenvironment on dermal fibroblast function. The Journal of Investigative Dermatology. 2013;**133**(5): 1362-1366

[96] Pittayapruek P, Meephansan J, Prapapan O, Komine M, Ohtsuki M. Role of matrix metalloproteinases in photoaging and photocarcinogenesis. International Journal of Molecular Sciences. 2016;**17**(6):E868

[97] Ke Y, Wang X-J. TGFβ signaling in photoaging and UV-induced skin cancer. The Journal of Investigative Dermatology. 2021;**141**(4):1104-1110

[98] Murphy G. Riding the metalloproteinase roller coaster. The Journal of Biological Chemistry. 2017;**292**(19):7708-7718

[99] Goldstein S, Moerman EJ, Fujii S, Sobel BE. Overexpression of plasminogen activator inhibitor type-1 in

senescent fibroblasts from normal subjects and those with Werner syndrome. Journal of Cellular Physiology. 1994;**161**(3):571-579

[100] Murano S, Thweatt R, Shmookler Reis RJ, Jones RA, Moerman EJ, Goldstein S. Diverse gene sequences are overexpressed in werner syndrome fibroblasts undergoing premature replicative senescence. Molecular and Cellular Biology. 1991;**11**(8):3905-3914

[101] Chojnowski A, Ong PF, Wong ES, Lim JS, Mutalif RA, Navasankari R, et al. Progerin reduces LAP2α-telomere association in Hutchinson-Gilford progeria. eLife. 2015;**4**:e07759

[102] Kortlever RM, Higgins PJ, Bernards R. Plasminogen activator inhibitor-1 is a critical downstream target of p53 in the induction of replicative senescence. Nature Cell Biology. 2006;**8**(8):877-884

[103] Krutmann J, Schikowski T, Morita A, Berneburg M. Environmentally-induced (extrinsic) skin aging: Exposomal factors and underlying mechanisms. The Journal of Investigative Dermatology. 2021;**141**(4S):1096-1103

[104] Yang JH, Lee HC, Wei YH. Photoageing-associated mitochondrial DNA length mutations in human skin. Archives of Dermatological Research. 1995;**287**(7):641-648

[105] Kaneko N, Vierkoetter A, Kraemer U, Sugiri D, Matsui M, Yamamoto A, et al. Mitochondrial common deletion mutation and extrinsic skin ageing in German and Japanese women. Experimental Dermatology. 2012;**21**(Suppl 1):26-30

[106] Berneburg M, Gremmel T, Kürten V, Schroeder P, Hertel I, von Mikecz A, et al. Creatine supplementation normalizes mutagenesis of mitochondrial DNA as well as functional consequences. The Journal of Investigative Dermatology. 2005;**125**(2):213-220

[107] Schroeder P, Gremmel T, Berneburg M, Krutmann J. Partial depletion of mitochondrial DNA from human skin fibroblasts induces a gene expression profile reminiscent of photoaged skin. The Journal of Investigative Dermatology. 2008;**128**(9):2297-2303

[108] Rizza ERH, DiGiovanna JJ, Khan SG, Tamura D, Jeskey JD, Kraemer KH. Xeroderma pigmentosum: A model for human premature aging. The Journal of Investigative Dermatology. 2021;**141**(4):976-984

[109] Hussain M, Krishnamurthy S, Patel J, Kim E, Baptiste BA, Croteau DL, et al. Skin abnormalities in disorders with DNA repair defects, premature aging, and mitochondrial dysfunction. The Journal of Investigative Dermatology. 2021;**141**(4):968-975

[110] Davis T, Wyllie FS, Rokicki MJ, Bagley MC, Kipling D. The role of cellular senescence in Werner syndrome. Annals of the New York Academy of Sciences. 2007;**1100**(1):455-469

[111] Lu H, Fang EF, Sykora P, Kulikowicz T, Zhang Y, Becker KG, et al. Senescence induced by RECQL4 dysfunction contributes to Rothmund-Thomson syndrome features in mice. Cell Death & Disease. 2014;**5**:e1226

[112] Chen Y-W, Harris RA, Hatahet Z, Chou K. Ablation of XP-V gene causes adipose tissue senescence and metabolic abnormalities. Proceedings of the National Academy of Sciences. 2015;**112**(33):E4556-E4564

[113] Le Clerc S, Taing L, Ezzedine K, Latreille J, Delaneau O, Labib T, et al. A genome-wide association study in

Caucasian women points out a putative role of the STXBP5L gene in facial photoaging. The Journal of Investigative Dermatology. 2013;**133**(4):929-935

[114] Chang ALS, Atzmon G, Bergman A, Brugmann S, Atwood SX, Chang HY, et al. Identification of genes promoting skin youthfulness by genome-wide association study. The Journal of Investigative Dermatology. 2014;**134**(3):651-657

[115] Xu J, Spitale RC, Guan L, Flynn RA, Torre EA, Li R, et al. Novel gene expression profile of women with intrinsic skin youthfulness by whole transcriptome sequencing. PLoS One. 2016;**11**(11):e0165913

[116] Law MH, Medland SE, Zhu G, Yazar S, Viñuela A, Wallace L, et al. Genome-wide association shows that pigmentation genes play a role in skin aging. The Journal of Investigative Dermatology. 2017;**137**(9):1887-1894

[117] Jacobs LC, Hamer MA, Gunn DA, Deelen J, Lall JS, van Heemst D, et al. A genome-wide association study identifies the skin color genes IRF4, MC1R, ASIP, and BNC2 influencing facial pigmented spots. The Journal of Investigative Dermatology. 2015;**135**(7):1735-1742

[118] Gao W, Tan J, Hüls A, Ding A, Liu Y, Matsui MS, et al. Genetic variants associated with skin aging in the Chinese Han population. Journal of Dermatological Science. 2017;**86**(1):21-29

[119] Liu F, Hamer MA, Deelen J, Lall JS, Jacobs L, van Heemst D, et al. The MC1R gene and youthful looks. Current Biology. 2016;**26**(9):1213-1220

[120] Chen B, Sun Y, Zhang J, Zhu Q, Yang Y, Niu X, et al. Human embryonic stem cell-derived exosomes promote pressure ulcer healing in aged mice by rejuvenating senescent endothelial cells. Stem Cell Research & Therapy. 2019;**10**(1):142

[121] Han J, Kraft P, Colditz GA, Wong J, Hunter DJ. Melanocortin 1 receptor variants and skin cancer risk. International Journal of Cancer. 2006; **119**(8):1976-1984

[122] Flood KS, Houston NA, Savage KT, Kimball AB. Genetic basis for skin youthfulness. Clinics in Dermatology. 2019;**37**(4):312-319

[123] Robinson MK, Binder RL, Griffiths CEM. Genomic-driven insights into changes in aging skin. Journal of Drugs in Dermatology (JDD). 2009; **8**(Suppl 7):s8-s11

[124] Lener T, Moll PR, Rinnerthaler M, Bauer J, Aberger F, Richter K. Expression profiling of aging in the human skin. Experimental Gerontology. 2006;**41**(4):387-397

[125] Makrantonaki E, Bekou V, Zouboulis CC. Genetics and skin aging. Dermatoendocrinology. 2012;**4**(3):280-284

[126] Kimball AB, Alora-Palli MB, Tamura M, Mullins LA, Soh C, Binder RL, et al. Age-induced and photoinduced changes in gene expression profiles in facial skin of Caucasian females across 6 decades of age. Journal of the American Academy of Dermatology. 2018;**78**(1):29-39.e7

[127] McGrath JA, Robinson MK, Binder RL. Skin differences based on age and chronicity of ultraviolet exposure: Results from a gene expression profiling study. The British Journal of Dermatology. 2012;**166**(Suppl 2):9-15

[128] Glass D, Viñuela A, Davies MN, Ramasamy A, Parts L, Knowles D, et al. Gene expression changes with age in skin, adipose tissue, blood and brain. Genome Biology. 2013;**14**(7):R75

[129] Makrantonaki E, Brink TC, Zampeli V, Elewa RM, Mlody B, Hossini AM, et al. Identification of biomarkers of human skin ageing in both genders. Wnt signalling—A label of skin ageing? PLoS One. 2012;**7**(11):e50393

[130] Wu H, Sun YE. Epigenetic regulation of stem cell differentiation. Pediatric Research. 2006;**59**(4):21-25

[131] Turner BM. Epigenetic responses to environmental change and their evolutionary implications. Philosophical Transactions of the Royal Society B: Biological Sciences. 2009;**364**(1534): 3403-3418

[132] Bormann F, Rodríguez-Paredes M, Hagemann S, Manchanda H, Kristof B, Gutekunst J, et al. Reduced DNA methylation patterning and transcriptional connectivity define human skin aging. Aging Cell. 2016;**15**(3):563-571

[133] Horvath S. DNA methylation age of human tissues and cell types. Genome Biology. 2013;**14**(10):3156

[134] Kabacik S, Horvath S, Cohen H, Raj K. Epigenetic ageing is distinct from senescence-mediated ageing and is not prevented by telomerase expression. Aging. 2018;**10**(10):2800-2815

[135] Wan DC, Wang KC. Long Noncoding RNA: Significance and potential in skin biology. Cold Spring Harbor Perspectives in Medicine. 2014;**4**(5):a015404

[136] Gerasymchuk M, Cherkasova V, Kovalchuk O, Kovalchuk I. The role of microRNAs in organismal and skin aging. International Journal of Molecular Sciences. 2020;**21**(15):5281

[137] Zhou B, Guo X, Zhang J, Xu Y, Li W, Wu D, et al. Elevated miR-34c-5p mediates dermal fibroblast senescence by ultraviolet irradiation. International Journal of Biological Sciences. 2013;**9**(7):743-752

[138] Abdelmohsen K, Gorospe M. Noncoding RNA control of cellular senescence. Wiley Interdisciplinary Reviews RNA. 2015;**6**(6):615-629

[139] Bannister AJ, Kouzarides T. Regulation of chromatin by histone modifications. Cell Research. 2011; **21**(3):381-395

[140] Orioli D, Dellambra E. Epigenetic regulation of skin cells in natural aging and premature aging diseases. Cell. 2018;**7**(12):268

[141] Contrepois K, Coudereau C, Benayoun BA, Schuler N, Roux P-F, Bischof O, et al. Histone variant H2A.J accumulates in senescent cells and promotes inflammatory gene expression. Nature Communications. 2017;**8**:14995

[142] Rübe CE, Bäumert C, Schuler N, Isermann A, Schmal Z, Glanemann M, et al. Human skin aging is associated with increased expression of the histone variant H2A.J in the epidermis. npj Aging and Mechanisms of Disease. 2021;**7**(1):1-11

[143] Narita M, Nũnez S, Heard E, Narita M, Lin AW, Hearn SA, et al. Rb-mediated heterochromatin formation and silencing of E2F target genes during cellular senescence. Cell. 2003;**113**(6):703-716

[144] Kosar M, Bartkova J, Hubackova S, Hodny Z, Lukas J, Bartek J. Senescence-associated heterochromatin foci are dispensable for cellular senescence, occur in a cell type- and insult-dependent manner and follow expression of p16(ink4a). Cell Cycle (Georgetown, Tex). 2011;**10**(3):457-468

[145] Distler JHW, Györfi A-H, Ramanujam M, Whitfield ML, Königshoff M, Lafyatis R. Shared and

distinct mechanisms of fibrosis. Nature Reviews Rheumatology. 2019;**15**(12): 705-730

[146] Wynn TA. Cellular and molecular mechanisms of fibrosis. The Journal of Pathology. 2008;**214**(2):199-210

[147] Shinde AV, Humeres C, Frangogiannis NG. The role of α-smooth muscle actin in fibroblast-mediated matrix contraction and remodeling. Biochimica et Biophysica Acta. 2017;**1863**(1):298-309

[148] Kendall RT, Feghali-Bostwick CA. Fibroblasts in fibrosis: Novel roles and mediators. Frontiers in Pharmacology. 2014;**5**:123

[149] Klingberg F, Hinz B, White ES. The myofibroblast matrix: Implications for tissue repair and fibrosis. The Journal of Pathology. 2013;**229**(2):298-309

[150] Jun J-I, Lau LF. The matricellular protein CCN1 induces fibroblast senescence and restricts fibrosis in cutaneous wound healing. Nature Cell Biology. 2010;**12**(7):676-685

[151] Babalola O, Mamalis A, Lev-Tov H, Jagdeo J. The role of microRNAs in skin fibrosis. Archives of Dermatological Research. 2013;**305**(9):763-776

[152] Auler M, Bergmeier V, Georgieva VS, Pitzler L, Frie C, Nüchel J, et al. miR-127-3p is an epigenetic activator of myofibroblast senescence situated within the microRNA-enriched Dlk1-Dio3–imprinted domain on mouse chromosome 12. Journal of Investigative Dermatology. 2021;**141**(4, Supplement):1076-1086.e3

[153] Yu Y, Schleich K, Yue B, Ji S, Lohneis P, Kemper K, et al. Targeting the senescence-overriding cooperative activity of structurally unrelated H3K9 demethylases in melanoma. Cancer Cell. 2018;**33**(2):322-336.e8

[154] Muñoz-Espín D, Serrano M. Cellular senescence: From physiology to pathology. Nature Reviews. Molecular Cell Biology. 2014;**15**(7):482-496

[155] Worm J, Bartkova J, Kirkin AF, Straten P, Zeuthen J, Bartek J, et al. Aberrant p27Kip1 promoter methylation in malignant melanoma. Oncogene. 2000;**19**(44):5111-5115

[156] Sarkar D, Leung EY, Baguley BC, Finlay GJ, Askarian-Amiri ME. Epigenetic regulation in human melanoma: Past and future. Epigenetics. 2015;**10**(2):103-121

[157] Rowdo FPM, Barón A, Gallagher SJ, Hersey P, Emran AA, Von Euw EM, et al. Epigenetic inhibitors eliminate senescent melanoma BRAFV600E cells that survive long-term BRAF inhibition. International Journal of Oncology. 2020;**56**(6):1429-1441

[158] Busam KJ, Pulitzer MP, Coit DC, Arcila M, Leng D, Jungbluth AA, et al. Reduced H3K27me3 expression in Merkel cell polyoma virus-positive tumors. Modern Pathology. 2017;**30**(6):877-883

[159] Gujar H, Mehta A, Li H-T, Tsai YC, Qiu X, Weisenberger DJ, et al. Characterizing DNA methylation signatures and their potential functional roles in Merkel cell carcinoma. Genome Medicine. 2021;**13**(1):130

[160] Park DE, Cheng J, McGrath JP, Lim MY, Cushman C, Swanson SK, et al. Merkel cell polyomavirus activates LSD1-mediated blockade of non-canonical BAF to regulate transformation and tumorigenesis. Nature Cell Biology. 2020;**22**(5):603-615

[161] Leiendecker L, Jung PS, Krecioch I, Neumann T, Schleiffer A, Mechtler K, et al. LSD1 inhibition induces differentiation and cell death in Merkel cell carcinoma. EMBO Molecular Medicine. 2020;**12**(11):e12525

[162] Elder SS, Emmerson E. Senescent cells and macrophages: Key players for regeneration? Open Biology. 2020;**10**(12):200309

[163] Krzyszczyk P, Schloss R, Palmer A, Berthiaume F. The role of macrophages in acute and chronic wound healing and interventions to promote pro-wound healing phenotypes. Frontiers in Physiology. 2018;**9**:419

[164] Davis FM, Gallagher KA. Epigenetic mechanisms in monocytes/macrophages regulate inflammation in cardiometabolic and vascular disease. Arteriosclerosis, Thrombosis, and Vascular Biology. 2019;**39**(4):623-634

[165] Zhuang Y, Lyga J. Inflammaging in skin and other tissues—The roles of complement system and macrophage. Inflammation & Allergy Drug Targets. 2014;**13**(3):153-161

[166] Korns D, Frasch SC, Fernandez-Boyanapalli R, Henson PM, Bratton DL. Modulation of macrophage efferocytosis in inflammation. Frontiers in Immunology. 2011;**2**:57

[167] Kale A, Sharma A, Stolzing A, Desprez P-Y, Campisi J. Role of immune cells in the removal of deleterious senescent cells. Immunity & Ageing. 2020;**17**(1):16

[168] Watt FM, Jensen KB. Epidermal stem cell diversity and quiescence. EMBO Molecular Medicine. 2009;**1**(5):260-267

[169] Voigt P, Tee W-W, Reinberg D. A double take on bivalent promoters. Genes & Development. 2013;**27**(12):1318-1338

[170] Bernstein BE, Mikkelsen TS, Xie X, Kamal M, Huebert DJ, Cuff J, et al. A bivalent chromatin structure marks key developmental genes in embryonic stem cells. Cell. 2006;**125**(2):315-326

[171] Lien W-H, Guo X, Polak L, Lawton LN, Young RA, Zheng D, et al. Genome-wide maps of histone modifications unwind in vivo chromatin states of the hair follicle lineage. Cell Stem Cell. 2011;**9**(3):219-232

[172] Takahashi K, Tanabe K, Ohnuki M, Narita M, Ichisaka T, Tomoda K, et al. Induction of pluripotent stem cells from adult human fibroblasts by defined factors. Cell. 2007;**131**(5):861-872

[173] Malik N, Rao MS. A review of the methods for human iPSC derivation. Methods in Molecular Biology (Clifton, N.J.). 2013;**997**:23-33

[174] Lapasset L, Milhavet O, Prieur A, Besnard E, Babled A, Aït-Hamou N, et al. Rejuvenating senescent and centenarian human cells by reprogramming through the pluripotent state. Genes & Development. 2011;**25**(21):2248-2253

[175] Lu Y, Qu H, Qi D, Xu W, Liu S, Jin X, et al. OCT4 maintains self-renewal and reverses senescence in human hair follicle mesenchymal stem cells through the downregulation of p21 by DNA methyltransferases. Stem Cell Research & Therapy. 2019;**10**(1):28

[176] Ocampo A, Reddy P, Martinez-Redondo P, Platero-Luengo A, Hatanaka F, Hishida T, et al. In vivo amelioration of age-associated hallmarks by partial reprogramming. Cell. 2016;**167**(7):1719-1733.e12

[177] Kurita M, Araoka T, Hishida T, O'Keefe DD, Takahashi Y, Sakamoto A, et al. In vivo reprogramming of wound-resident cells generates skin epithelial tissue. Nature. 2018;**561**(7722):243-247

[178] Sarkar TJ, Quarta M, Mukherjee S, Colville A, Paine P, Doan L, et al. Transient non-integrative expression of nuclear reprogramming factors promotes multifaceted amelioration of aging in human cells. Nature Communications. 2020;**11**(1):1-12

Chapter 4

Cellular Senescence in Bone

Danielle Wang and Haitao Wang

Abstract

Senescence is an irreversible cell-cycle arrest process induced by environmental, genetic, and epigenetic factors. An accumulation of senescent cells in bone results in age-related disorders, and one of the common problems is osteoporosis. Deciphering the basic mechanisms contributing to the chronic ailments of aging may uncover new avenues for targeted treatment. This review focuses on the mechanisms and the most relevant research advancements in skeletal cellular senescence. To identify new options for the treatment or prevention of age-related chronic diseases, researchers have targeted hallmarks of aging, including telomere attrition, genomic instability, cellular senescence, and epigenetic alterations. First, this chapter provides an overview of the fundamentals of bone tissue, the causes of skeletal involution, and the role of cellular senescence in bone and bone diseases such as osteoporosis. Next, this review will discuss the utilization of pharmacological interventions in aging tissues and, more specifically, highlight the role of senescent cells to identify the most effective and safe strategies.

Keywords: bone, senescence, osteoporosis, bone remodeling, telomere dysfunction, senolytic drug

1. Introduction

Aging is an inevitable physiological condition that comes with organ and tissue function impairment. It is the most significant risk factor for developing chronic diseases, including cancer, cardiovascular disease, metabolic dysfunction, osteoarthritis, and osteoporosis. Osteoporosis originated from the Greek word for porous bones, is one of the most common metabolic diseases. Associated with advancing chronological age, it affects more than 200 million patients worldwide and increases morbidity, mortality, and creates a significant burden of economic expenditures [1, 2]. Given that the population segment with the most rapid growth is the elderly in many countries, osteoporosis could present a global challenge impacting affected individuals' health quality and life span. Characteristics of aging bone are low bone mineral density and deterioration of bone architecture, producing weakened bone prone to fractures. Thus, osteoporosis presents severe global health concerns, disposing to over 9 million fractures every year [3]. Senescent cells play a crucial role in aging bone; therefore, it is essential to understand the cellular and molecular mechanisms to develop treatments to prevent age-related diseases and maximize a healthy life span. This chapter provides a comprehensive treatise of senescence in bone and emerging therapeutic approaches to treatment.

2. Physiology of bone tissue

The skeletal system is one of the most complex structures in mammals and is essential for storing and maintaining the homeostasis of the body's minerals. Composed of various bones, cartilages, ligaments, tendons, and other tissues, it provides the framework for the body, supports locomotion, and protects vital organs such as the brain and bone marrow. It is commonly thought that the metabolic functions are carried out primarily by trabecular bone and the mechanical functions mainly by cortical bone. Bone, specifically, is a complex tissue that exhibits four types of cells: osteoclasts, bone lining cells, osteocytes, and osteoblasts. In addition, it houses bone marrow and serves as the main reservoir for the body's calcium and phosphate.

Bone is a highly dynamic tissue that adapts to change and is constantly shifting throughout life. The most rapid rate of bone modeling occurs during childhood and adolescence, where bones are architecturally modified to support skeletal functions. Moreover, human skeletal tissue is in a constant state of remodeling throughout life [4]. A retained net bone mass is needed for homeostasis.

3. Senescence in bone

Discovered more than five decades ago by Hayflick and Moorhead [5], cellular senescence has played a significant role in our understanding and advancement in science. By definition, cellular senescence is a permanent state of cell cycle arrest characterized by specific phenotypic changes [6]. Characteristics include distinct cellular morphological alterations, gene expression, chromatin structure, cell signaling, and the senescence-associated secretory phenotype (SASP). Cellular senescence is found in bone and promotes age-related diseases such as osteoporosis [7]. In addition, senescent cells damage bone remodeling by impairing bone formation and osteoblast progenitor cell function, thus promoting osteoclastogenesis [8]. This is triggered by various stressors, including oxidative stress, genomic instability, and telomere shortening (replicative senescence). Telomeres protect chromatins and help maintain replication and genome stability.

The various physiological and pathological processes such as remodeling, aging, and injury can cause cells to become senescent. With aging, more cells become senescent and accumulate in tissues, including bones. A prominent characteristic of cell senescence is the SASPs, which are proinflammatory proteins that are primary contributors to their disease-inducing properties. Cyclin-dependent kinase inhibitors (CDKis) such as p16, p21, p27, the release of cytokines, chemokines, and soluble factors, causes this impaired microenvironment known as SASP. The SASP increases proinflammatory factors and upregulates NF-κB, contributing to aging bone disease [9].

As a hallmark of aging, it is essential to understand cellular senescence to effectively identify novel drugs to treat osteoporosis. Moreover, targeting cellular senescence has emerged as a therapeutic target for preventing or treating age-related diseases. Clearing these cells in mouse models has delayed tissue and organ dysfunctions [10]. In addition, senescence has been shown to have antiproliferative effects, a fundamental key to identifying novel drugs to treat osteoporosis.

4. Etiology of bone senescence

Bone loss is a part of the natural aging process in both men and women [11]. Developmental, genetic, and lifestyle factors (lack of physical activity, injuries,

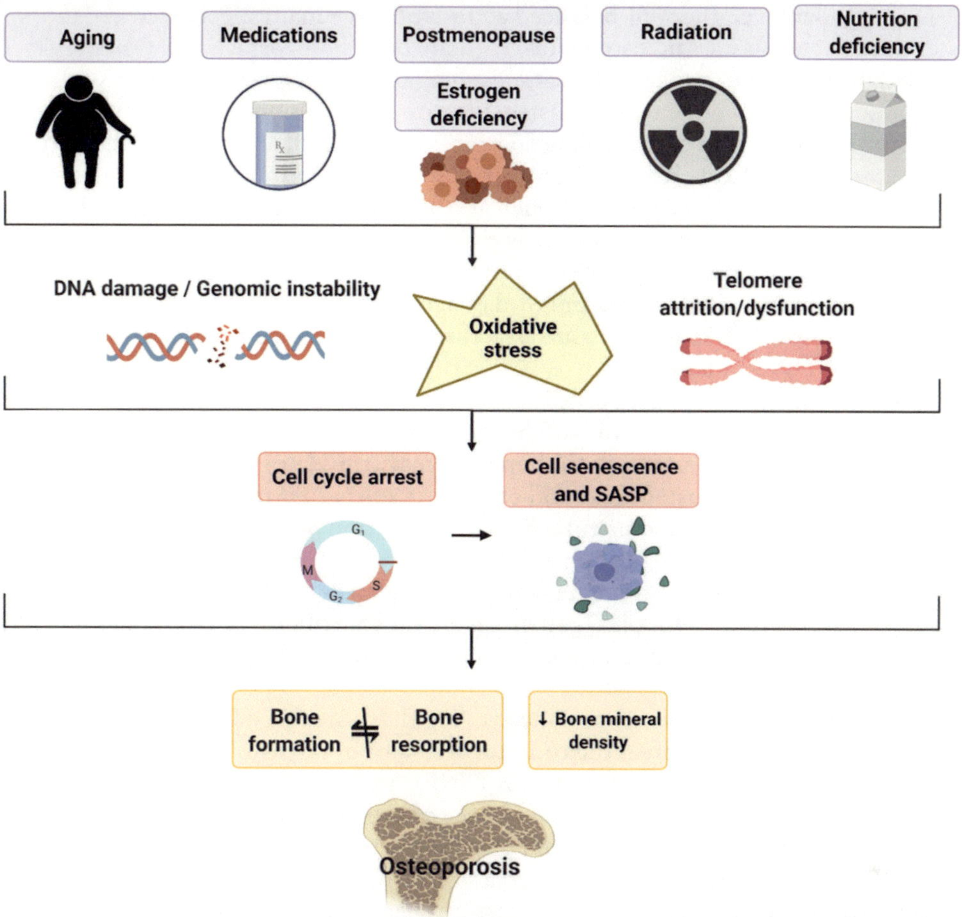

Figure 1.
Pathogenesis of osteoporosis. Aging and various environmental exposures can induce DNA damage and instability, oxidative stress, telomere attrition, dysfunction at the molecular level, and cell cycle arrest and senescence at the cellular level. These will break the remodeling process of bone formation and resorption, decrease bone mineral density, and progress to osteoporosis.

medication use, smoking, poor diet) contribute to bone fragility in older people. The skeletal system goes through progressive bone loss, where changes in bone quality and quantity will occur. An accumulation of weakened skeletal bone may result in osteoporosis. Advancing chronological age is one of the significant risk factors for osteoporosis [12]. Characteristics of aging bone include low bone mineral density and weakened bone architecture, significantly increasing the risk of fractures for affected individuals (**Figure 1**).

5. Osteocyte and osteoblast differentiation

Throughout life, old bone is replaced by new bone, a process termed bone remodeling. This continuous cycle is necessary for fracture healing and adaptation to mechanical strains such as exercise. Bone regeneration occurs within bone cavities to target and replaces bone with accumulated microfracture fatigue.

On a cellular level, the well-balanced actions of three main specialized cell types, osteoclasts, osteoblasts, and osteocytes, regulate bone homeostasis [4]. Osteoclasts resorb damaged bone, and osteoblasts subsequently refill the resorbed area with an equal amount of new bone matrix. Osteocytes are mechanosensory cells that act as the central coordinators of this balanced process in transmitting signals needed to sustain mechanical loads [13, 14]. Disruption among the actions of this repertoire can turn to bone pathological conditions such as osteoporosis and rheumatoid arthritis. On a subcellular level, the bone matrix is changed by rearrangement of trabecular struts, changes in calcium deposition, subperiosteal expansion, and enlargement of the medullary cavity. Unrepaired micro-damaged bone reduces bone health, resulting in the mechanical failure of the tissue (fracture). The remodeling process is the same in cortical and trabecular bone.

Under normal physiological conditions, the amount of bone resorbed and replaced is equal, maintaining the bone mass. This process relies on having an adequate supply of osteoblasts, which comes from the generation of stimulatory signals for osteoblast formation produced by osteoclasts and osteocytes released during resorption [15]. Osteocytes regulate this fundamental bone regeneration process by sending signals to osteoclasts and osteoblasts to control their actions [16]. Furthermore, there is an association between lower osteocyte density in human central cancellous bone and increased surface remodeling [17], an independent contributor to bone fragility [18]. Therefore, a primary strategy in finding therapeutic targets to treat osteoporosis involves targeting osteoclasts [19].

Several molecular mechanisms concur to regulate osteoblast/osteoclast/osteocyte activity. The main one involves the receptor activator of nuclear factor-kappa-B ligand (RANKL) of tumor necrosis factor (TNF) superfamily ligand 11 (TNFSF11) [20]. This cytokine is expressed on the surface of osteoblasts and osteocytes. On the membrane of osteoclast precursors and mature osteoclasts, RANKL binds to its receptor RANK, a ligand-receptor binding process termed the critical paracrine system, regulating osteoclast function. This process can be inhibited by osteoprotegerin (OPG), a decoy of RANKL produced by osteoblasts and osteocytes.

Moreover, osteocytes regulate bone formation by secreting modulators of the wingless-type mouse mammary tumor virus [MMTV] integration site members (Wnt) signaling pathways. These include activators nitric oxide and ATP, inhibitors sclerostin SOST, as well as dickkopf-related protein 1 (DKK1)). Wnts modulate cell proliferation, differentiation, and stem cell remodeling [21]. Previous studies have found that the activation of Wnts impacts osteoblasts and osteoblast lineages by increasing quantities and enhancing the functionality of osteoblasts [22]. Recently, studies were done in vivo to test whether the Fzd-Lrp receptor with Wnt mimetics can activate Wnt/β-catenin signaling and promote rapid bone growth [23]. It was found that within 2 weeks after treatment with selected Wnt mimetics, bone mineral density and vertebral cortical and trabecular bone growth increased significantly [23]. This could provide a therapeutic therapy used to target bone diseases such as osteoporosis.

However, with aging, the bone remodeling process is affected. Osteoporosis occurs when bone metabolism is perturbed. In addition, chronic diseases such as estrogen deficiency, malignant disease, and chronic inflammation also cause the uncoupling of osteoclasts and osteoblasts [24, 25]. As a result, less new bone is formed relative to the resorption of old bone, ending in a net bone loss. The cortical and trabecular thinning thereby leads to an overall bone loss and fragility. Thus, bone remodeling causes a drastic loss of bone mass and strength over prolonged periods, eventually osteoporosis.

6. Sex steroid deficiency

The process of senescence in bone begins after peak bone mass is reached. This is generally during the third decade of life but varies between sexes. Estrogens and androgens are hormones that play crucial roles in skeletal homeostasis during growth and adulthood.

Estrogen is the primary hormonal regulator of bone metabolism, inhibiting osteoblast and osteocyte apoptosis [26, 27]. Therefore, a decrease in androgen and estrogen levels negatively affects bone remodeling by causing the uncoupling of osteoclasts and osteoblasts [28]. Hormonal withdrawal also contributes to mineral disturbances with calcium absorption [29].

The association between a decline of estrogen levels in postmenopausal women and the onset of osteoporosis was first noted by Fuller Albright in 1940. Since then, estrogen deficiency has become the primary cause of bone loss in older women [11]. An accelerated decrease occurs in the perimenopausal period when there is rapid bone remodeling. As a result, women experience the loss of whole trabecular components and combined with a negative remodeling balance, the bone loses mass and strength. In addition, estrogen levels affect T cells by increasing tumor necrosis factor secretion, promoting RANKL-induced osteoclastogenesis [30]. In men, a loss of both estrogens and androgens is associated with a loss of bone mass and the development of osteoporosis [31]. Small increases in estrogen levels can improve bone health without some of the adverse effects of conventional-dose estrogen therapy [32]. Sex steroids can regulate osteoclastogenesis and the survival of osteoclasts [33].

7. Pathology

Cellular senescence has been identified as a response to multiple stressors. Common denominators of aging include telomere attrition, genotoxic agents, oxidative stress, chromosome instability, and oncogene activation. Skeletal involution results from the accumulation of poor nutrition, immobility, and the effects of treatments, all of which often come with old age. Mediated with bone remodeling, the progressive and cumulative pathologies of these factors contribute to the pathogenesis of osteoporosis.

7.1 SASP

Bone homeostasis is a balanced equilibrium between osteoblast and osteoclast activities. In senescent cell microenvironments, osteocytes control myeloid lineage cells [34]. Therefore, the SASP can be the cause of some of the severe effects of senescent cells. With aging, more osteocytes become senescent that acquire a new phenotype. As a result, they secrete various factors, including proinflammatory cytokines, growth modulators, which collectively comprise the SASP. Regulated at epigenetic, transcriptional, and posttranscriptional levels, SASP plays a critical role in contributing to various outputs of senescence [35]. For example, SASP factors mediate developmental senescence, wound healing, and tissue plasticity. In addition, the SASPs secrete signals that are communicated and amplified by neighboring myeloid lineage cells (such as B cells, osteoblasts, and T cells), resulting in the overproduction of proinflammatory cytokines. As a result, it contributes to chronic inflammation and creates a toxic local microenvironment that contributes to age-related bone loss.

7.2 DNA damage

DNA damage is considered to be the root of aging-associated multimorbidity [36]. It is caused by exposure to harmful exogenous factors (such as chemical compounds in the environment, chemotherapy, and UV radiation from the sun) and endogenous factors (such as reactive oxygen species and metabolic by-products). Consequences of accumulated DNA damage happen on the cellular and molecular levels. With aging, there are impaired cell and organ functions, inflammation, and cancer [36]. On the cellular level, DNA damage induces permanent cell-cycle arrest. It molecularly triggers genome instability with chromosome aberrations and mutations. Irreparable DNA damage accumulation in tissues and organs leads to cellular senescence, one of the main driving forces of aging [37].

7.3 Telomeres

Telomere dysfunction is induced in response to DNA damage. About half of the DNA damage foci in senescent cells localize to telomeres. Accumulated and progressive telomere shortening is a senescence biomarker and drives the aging process, a concept first discovered in the late 1980s [38]. Telomeres are short DNA sequences found at the ends of eukaryotic chromosomes that determine cellular life span [39]. Telomeric TTAGGG repeats and compound proteins make up the ends of chromosomes or the cap. The cap protects the telomere ends from appearing as double-break strands and prevents chromosome fusion and genome degradation.

During each cell replication, DNA polymerase cannot fully replicate chromosome ends, resulting in a loss of DNA. Accumulation of DNA damage at telomeres causes uncapping. With each cell division, telomeres shorten in length, and cell proliferation is restricted [40], a phenomenon termed replicative senescence [41]. To counteract telomere shortening, a specialized ribonucleoprotein enzyme called telomerase synthesizes new telomeric DNA [42].

The result of telomere shortening is telomeric DNA loop destabilization and telomere uncapping, which produces telomere dysfunction-induced foci (TIFs). This further activates the DNA damage repair (DDR), which recognizes double-strand breaks and activates the p53/p21 and p16 pathways [43]. These factors result in the pre-senescent cells withdrawing from the cell cycle and becoming senescent, which increases with age [44]. Furthermore, through inflammatory cytokines and impaired growth signaling, DDR results in replicative senescence [43].

7.4 Oxidative stress/ROS

Oxidative stress is a potential cause of results from various diseases and an important mechanism in bone degradation. Aging causes an increase in reactive oxygen species (ROS), which results in an imbalance of ROS and antioxidant defenses. Increased reactive oxygen species influence numerous cellular processes, including the timing of death by apoptosis, and have been linked to aging and the development of age-related diseases. It can damage DNA and contribute to aging. It has been found that oxidative stress increases with age in the bone of female or male C57BL/6 mice [33]. Oxidative stress alters the bone remodeling process by disrupting osteoclast and osteoblast activity. This can result in low bone mineral density, the characteristics of osteoporosis.

7.5 Oncogene stress

DNA damage is also responsible for oncogene-induced senescence (OIS). Oncogenic stress is commonly known as a critical mechanism of cancer. Oncogene activation is genetic stress and phenotypic changes that induce senescence. With activated oncogenes, there are high levels of replication. Pathways such as the ataxia telangiectasia and Rad3-related (ATR), ataxia–telangiectasia mutated (ATM), and p53 converge with the cyclin-dependent kinase inhibitors p16, p21, and p27 and hyperphosphorylation of the retinoblastoma protein, thereby triggering withdrawal from the cell cycle [45].

7.6 Glucocorticoid-induced osteoporosis

Glucocorticoids are drugs used to suppress allergic, autoimmune, and inflammatory diseases. However, prolonged use of glucocorticoids can result in complications such as glucocorticoid-induced osteoporosis (GIO). Glucocorticoids cause senescence in various cell lines and have been found to stimulate the p21 gene expression. During the initial treatment, this drug increases bone resorption with an enhancement of osteoclast maturation and differentiation. However, long-term use inhibits osteoclastogenesis by promoting apoptosis of osteoblasts and osteocytes [46, 47]. Dexamethasone, a type of glucocorticoid, was found to promote cell senescence and activate parts of SASP through inhibition of osteoblast function [48]. This resulted in decreased bone formation and increased bone resorption. Other effects include suppressing insulin-like growth factor 1, which promotes bone formation and further causes collagen degradation and osteoblast apoptosis [48].

7.7 Chronic inflammatory diseases

Chronic inflammatory diseases are associated with bone loss, which increases bone resorption and decreases bone formation, resulting in a bone deficit [24, 49].

The summary of the pathological factors that induce cellular senescence is provided in **Table 1**.

Pathological factors	Causes	Mechanisms
SASP	Aging	Chronic inflammation
DNA damage	Aging, environmental factors	Cellular senescence
Telomere dysfunction	DNA damage	Telomere uncapping, activates the p53/p21 and p16 pathways
Oxidative stress / ROS	Aging	imbalance of ROS and antioxidant defenses
Oncogene stress	DNA damage (i.e. cancer)	oncogene-induced senescence. Inhibits, osteoclastogenesis
Glucocorticoid	Glucocorticoid drugs	stimulate the p21 gene expression, Osteoblast apoptosis
Chronic-inflammation	Chronic-inflammatory diseases (i.e. arthritis	Increase osteoclast function, decrease osteoblast function

Table 1.
Summarized mechanisms of bone senescence.

8. Treatment

Both nonpharmacological (lifestyle factors, supplements) and pharmacological (antiresorptive and anabolic) treatments exist. The chapter also highlights the ongoing advancements of senescence research on aging-bone diseases (**Figure 2**).

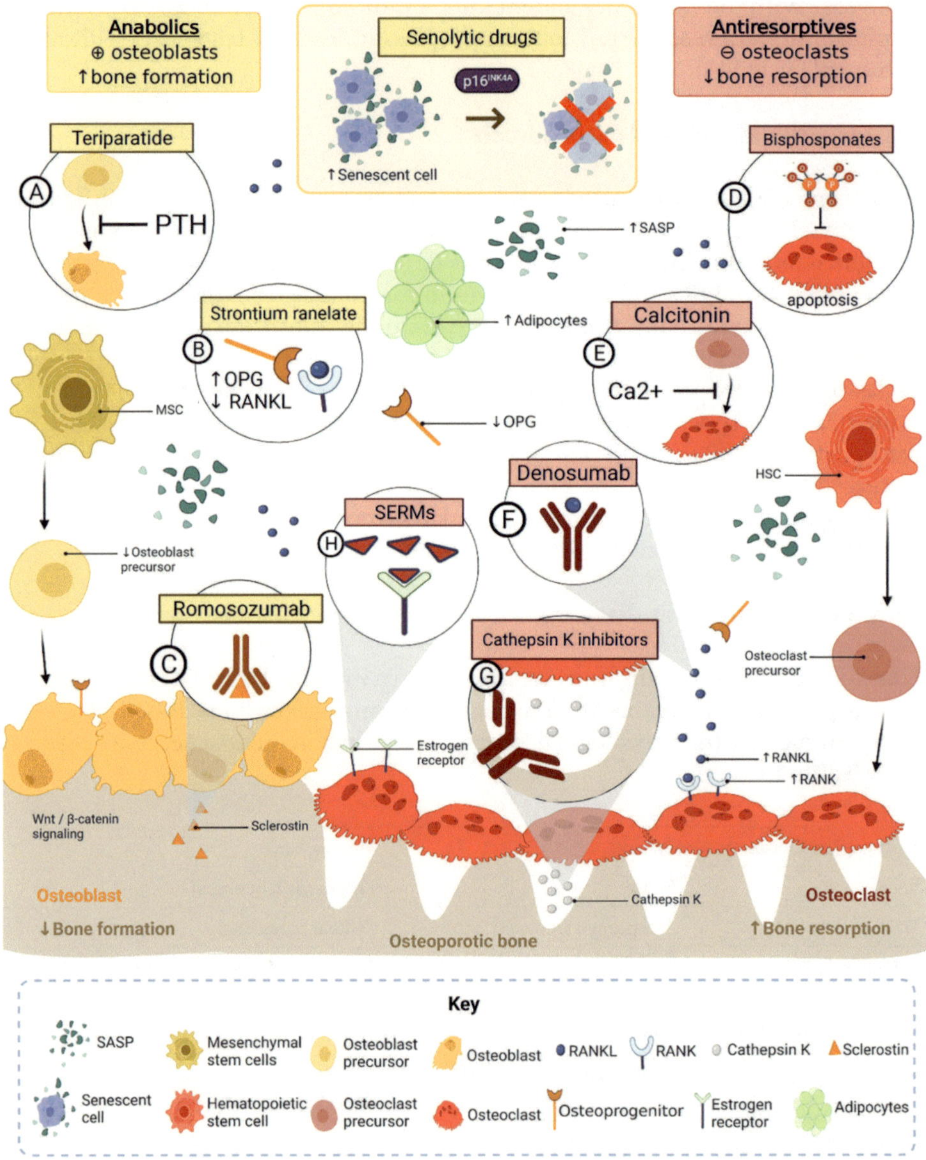

Figure 2.
Treatments of osteoporosis. Antiresorptive and anabolic and senolytic treatments of osteoporosis. These treatments target different pathways. Anabolic treatment options include teriparatide (A), strontium ranelate (B), and romosozumab (C). The antiresorptive treatment includes bisphosphonates (D), calcitonin (E), denosumab (F), cathepsin K inhibitors (G), and SERMs (H). With particular regard, senolytic drugs treatment includes Fisetin, Dasatinib, quercetin, and D + Q.

9. Nonpharmaceutical

Optimizing lifestyle factors by diet and physical exercise is beneficial to bone health. Physical exercise and an active lifestyle have a significant impact on bone health. During the muscular activity, the mechanical forces produced are sensed by osteocytes and promote bone growth. In response to exercise, skeletal muscle also secretes myokines, which are molecules that directly affect bone metabolisms, such as irisin, myostatin, and insulin-like growth factor-1 [50]. Exercise also restores body coordination and balance, decreasing the risk of falls, especially among older people. On the other hand, limited physical movement and muscle atrophy with old age result in osteoporosis [51].

In addition, an increase in nutrient intake, specifically vitamin D, protein, calcium, and vitamin K2, will slow osteoporotic regression. Vitamin D has a critical role in regulating calcium homeostasis and bone metabolism. In addition, calcium and vitamin D can suppress serum levels of parathyroid hormones and stimulate bone growth, making them have an antiresorptive effect. The daily calcium intake recommendation is between 800 and 1200 mg, and vitamin D intake is 800 IU per day for men and women over 50 [52].

Vitamin D insufficiency and low serum calcium levels are widespread in elderly people, contributing to lower BMD and increased bone fragility [53]. Dietary sources are the preferred option, but supplementation is beneficial, especially in elderly people. With daily calcium and vitamin D supplements, fracture risk drops significantly, making them essential in aging-bone disease treatments [54]. In most clinical studies testing the efficacy of antiresorptive and anabolic therapies, calcium and vitamin D have been used. When given together, they have been found to have been effective in preventing fractures [53, 55]. However, in most clinical cases, calcium supplementation is subsidiary to bisphosphonates or anti-RANKL drugs [56].

Nutraceuticals are substances including isolated nutrients, dietary supplements, herbal products, and medical foods. For example, higher intakes of antioxidants, phytoestrogens (plant compounds that function like estrogen agonist-antagonists), and other minerals such as phosphorus can be markers for a healthy lifestyle [57, 58]. Phosphorus is another critical factor in preventing aging-bone diseases such as osteoporosis. It is an essential nutrient for bone formation, but too much of it harms bone health [58].

Physical exercise and muscle fitness have a dramatic impact on bone health. Muscle secretes a set of molecules, known as myokines, directly affecting bone metabolisms, such as irisin, myostatin, and insulin-like growth factor-1. During activity that produces mechanical force, osteocytes sense this and convert it into bone deposition. On the contrary, disuse or muscle atrophy results in osteoporosis.

10. Pharmaceutics

The search for armamentariums targeting metabolic bone diseases is increasing. Currently, various antiresorptive and anabolic therapies are available as treatments for osteoporosis [23]. Antiresorptive therapies are the most common pharmacological tools to prevent osteoporosis progression. These drugs inhibit osteoclast proliferation and the recruitment and differentiation of its precursors [54]. It is suggested for early menopausal women or patients with moderate osteoporosis. Anabolic therapies are another option for treatment that targets osteoblasts to stimulate bone mineralization.

In comparison to antiresorptive medications, anabolic agents reduce fracture risk more efficiently. Thus, these should be considered first-line therapy for patients at very high risk or with a history of vertebral fracture [59]. In addition, pharmaceutical medications seek to improve bone fidelity and architectural foundation for long-term skeletal health. Therefore, the search for armamentariums targeting skeletal diseases is increasing. Currently, various antiresorptive and anabolic therapies are available as treatments for osteoporosis [23].

Bone homeostasis is a balanced equilibrium between osteoblast and osteoclast activities. In senescent cell microenvironments, osteocytes control myeloid lineage cells [34]. With aging, more osteocytes become senescent that produces SASP signals. These signals are communicated and amplified by neighboring myeloid lineage cells (such as B cells, osteoblasts, and T cells), resulting in the overproduction of proinflammatory cytokines. As a result, a toxic local microenvironment is created that contributes to age-related bone loss.

The antioxidant NAC, coupled with estrogens or androgens in male and female mice, prevents a gonadectomy-induced increase in oxidative stress, bone loss, osteoblast, and osteocyte apoptosis. So, sex steroids can regulate osteoclastogenesis and the survival of osteoclasts via antioxidant actions [33].

10.1 Antiresorptive

Antiresorptive therapy is the most common pharmacological tool to prevent osteoporosis progression. These drugs inhibit osteoclasts' proliferation and the recruitment and differentiation of their precursors [54].

Bisphosphonates (BPs) are the primary therapeutic options used to inhibit osteoclast-mediated bone resorption. These nitrogen-containing drugs have a strong affinity for bone apatite in vitro and in vivo. BPs bind to hydroxyapatite crystals on bone surfaces and inhibit the mevalonate pathway in osteoclasts, increasing apoptosis. This preferentially occurs in sites with accelerated skeletal turnover rates. BPs have been shown to increase bone mineral density (BMD), reduce bone turnover markers, and reduce the risk of osteoporotic fractures. Some drug options include alendronate, risedronate, and zoledronic acid. Currently, they are the most common and effective drugs used for osteoporosis, Paget's disease, and inflammation-related bone loss [60].

Denosumab, an anti-RANKL antibody, is a fully human monoclonal antibody to the RANKL, which blocks its binding to RANK. The prevention of RANKL and its receptor RANK interaction thereby inhibits osteoclast differentiation [61]. Presently, denosumab is the only FDA-approved monoclonal antibody to treat osteoporosis. These antiresorptive agents have been most influential in decreasing the risk of vertebral fractures by more than 50%, nonvertebral fractures by 20–25%, and hip fractures by 40–50% [62].

Selective Estrogen Receptor Modulators (SERMs) are an alternative for estrogen and are used primarily in postmenopausal women of younger age. SERMs rely on their tissue-selective estrogen receptor agonist or antagonist activity and their interaction with the estrogen receptor. They interact with the RANKL/RANK/OPG system and downmodulate osteoclast function [63]. This process allows for the treatment of vasomotor systems and the prevention of osteoporosis [64]. Various SERMs, including raloxifene, which represents dual agonistic and antagonistic properties in estrogenic pathways, have decreased bone fragility. In postmenopausal women with low BMD, raloxifene has been shown to reduce vertebral fracture risk by 30–50% [63].

In particular, this drug is recommended for patients with a family history of breast cancer, as it has also significantly demonstrated reduced risks of breast cancer in women [29].

Calcitonin receptors are found on osteoclasts and osteoblasts and serve as regulators of osteoclast function and maturation. Calcitonin is a naturally occurring peptide hormone that binds to specific receptors primarily on the surface of osteoclasts to inhibit bone resorption activity strongly. It has been used to treat osteoporosis for many years, especially for patients with acute osteoporotic fractures and postmenopausal women [65].

Cathepsin K (CatK) is one of the most potent proteases in the lysosomal cysteine proteases family. CatK's primary function is to mediate bone resorption, making it a strategic target for osteoporosis treatments. The only CatK inhibitor candidate, Odanacatib (ODN), was developed by Merck & Co. Phase III clinical trials; it showed high therapeutic efficacy in patients with postmenopausal osteoporosis but was terminated due to the cardio-cerebrovascular adverse effects. As of now, there is no available drug approved by the FDA that targets cathepsin k but is an ongoing direction for osteoporosis treatment [66].

10.2 Anabolics

PTHrP is required for normal bone development. Teriparatide is a bioactive form of the parathyroid hormone of recombinant human PTH 1–34 fragment rhPTH (1–34) [67]. It is the first and only available therapeutic agent that activates and stimulates osteoblasts. In contrast with antiresorptive therapy, teriparatide increases bone formation by inhibiting sclerostin production in osteocytes and increases bone resorption by stimulating RANKL production by osteoblasts and osteocytes. In addition, PTH inhibits p16Ink4a and thereby downregulates senescence [68]. Intermittent administration of PTH increases osteoblast amounts and activities, thereby improving skeletal architecture at both trabecular and cortical bone sites [69].

Furthermore, this drug provides some remediation of the architectural defects in the osteoporotic skeleton [70]. Daily injections of teriparatide in patients with severe osteoporosis can reduce hip fractures by 56% [71]. Abaloparatide, a 34 amino acid synthetic analog of parathyroid hormone-related protein analog drug, is an FDA-approved drug to treat postmenopausal osteoporosis.

Wnt signaling pathways modulate cell proliferation, differentiation, and stem cell remodeling [21]. Activation of Wnts impacts bone remodeling by increasing quantities and enhancing the functionality of osteoblasts. The discovery of this pathway has opened the way to new anabolic treatments. For example, sclerostin is a protein secreted primarily by osteocytes and protects against the excessive bone formation. Anti-sclerostin antibodies stimulate osteoprotegerin production, leading to decreased bone resorption and uncoupling of osteoclast and osteoblast activity [4]. In addition, romosozumab, an anti-sclerostin monoclonal antibody that binds sclerostin, has favorable dual effects on bone by increasing bone formation and reducing bone resorption [72]. In studies done with postmenopausal women prone to osteoporosis, a dose of 210 mg romosozumab monthly amounts resulted in significantly increased BMD and was more effective than daily teriparatide or weekly alendronate doses [73]. Thus, it is considered another emerging therapeutic for skeletal aging.

Strontium ranelate is a relatively novel drug currently approved in Europe for the treatment of postmenopausal osteoporosis. It has dual effects of inhibiting bone resorption and promoting bone formation [74, 75]. It can stimulate the differentiation of pre-osteoblasts into osteoblasts and promotes osteoblast release of OPG. This can act as a decoy receptor for RANKL and thereby interfere with osteoclast differentiation. In every gram of bone, strontium is naturally occurring in trace amounts at around 100 μg. In other words, the therapeutic strategy with strontium ranelate is producing more strontium available to incorporate into bone [76]. In other words, the therapeutic approach with strontium ranelate is producing more strontium available to incorporate into bone [76].

Dual acting treatments that can coordinately stimulate osteoblasts and inhibit osteoclasts have significantly improved bone quality compared with monotherapy [77]. For example, a combination of teriparatide and denosumab generated more significant increases in BMD and bone strength than independent use of either drug [77]. In addition, the combination of Wnt mimetics and current clinical treatments has been found to improve bone mass and strength [23]. Thus, compared with monotherapy, sequential therapy can improve bone health and serve as an emerging option for treatment. Compared with monotherapy, dual-acting treatments that can coordinately stimulate osteoblasts and inhibit osteoclasts have significantly improved bone quality [67]. For example, the combination of Wnt mimetics and current clinical treatments has been found to improve bone mass and strength [23]. Thus, in comparison to monotherapy, sequential therapy has the potential to improve bone health significantly.

10.3 Senolytic drugs

Interest in targeting senescence to halt or prevent age-related diseases, also known as senotherapy, has grown. Senolytic drugs are SASP modulators that eliminate cell senescence. More cells become senescent with advancing age and accumulate in tissues, suggesting that targeting the senescent cells is a promising treatment. Hence, several studies have explored senescent cells and their role in aging-bone diseases. The first thorough evidence showing senescence in mammalian bone cells was found in 2016 [78]. Osteocytes have the vital role of orchestrating bone remodeling, and osteocytes with senescence attributes contribute to osteoporosis [78]. To build off of this, another study found that genetically eliminating senescent cells and their SASP could prevent age-related osteoporosis [79]. In addition, the elimination of p16Inka-senescent cells improved bone quality. To build on this finding, researchers performed another study and found that genetically eliminating senescent cells and their SASP could prevent age-related osteoporosis [79]. Also, the elimination of p16Inka-senescent cells improved bone quality [10]. In mice, senolytic intervention improved bone mass, strength, and microarchitecture [7]. Novel drugs that use this strategy include Dasatinib (D), Quercetin (Q), D + Q [80], and Fisetin [81]. Senolytic drugs have shown a positive impact on bone metabolism by preventing bone loss and increasing health span.

11. Conclusion

The cellular morphological changes that come with aging dramatically affect bone health and increase the risk of developing age-related bone diseases. The sequelae of osteoporosis include decreased bone mass and increased pronation to fractures, a significant concern for the aging population. Recent literature is addressing utilizing new pharmaceutical targets to reverse or treat the adverse effects of aging.

For example, cell senescence in bone paves the way for developing new therapeutic targets. With improved knowledge of the pathophysiology of osteoporosis and new targets, potential new treatments are being investigated. The use of pharmaceuticals and nonpharmaceuticals appears promising in preventing or treating aging bone diseases, including osteoporosis.

Author details

Danielle Wang and Haitao Wang*
Division of Geriatric and Medicine and Gerontology, Departments of Internal Medicine, Department of Physiology and Biomedical Engineering, Mayo Clinic, Rochester, Minnesota, USA

*Address all correspondence to: wang.haitao@mayo.edu

References

[1] Lane NE. Epidemiology, etiology, and diagnosis of osteoporosis. American Journal of Obstetrics and Gynecology. 2006;**194**(2):S3, S11

[2] Ström O et al. Osteoporosis: burden, health care provision and opportunities in the EU: a report prepared in collaboration with the International Osteoporosis Foundation (IOF) and the European Federation of Pharmaceutical Industry Associations (EFPIA). Archives of Osteoporosis. 2011;**6**:59, 155

[3] Johnell O, Kanis JA. An estimate of the worldwide prevalence and disability associated with osteoporotic fractures. Osteoporosis International. 2006;**17**(12): 1726, 33

[4] Baron R, Hesse E. Update on bone anabolics in osteoporosis treatment: rationale, current status, and perspectives. The Journal of Clinical Endocrinology and Metabolism. 2012;**97**(2):311, 25

[5] Hayflick L, Moorhead PS. The serial cultivation of human diploid cell strains. Experimental Cell Research. 1961;**25**(3): 585, 621

[6] López-Otín C et al. The hallmarks of aging. Cell. 2013;**153**(6):1194, 217

[7] Farr JN et al. Targeting cellular senescence prevents age-related bone loss in mice. Nature Medicine. 2017; **23**(9):1072, 1079

[8] Childs BG et al. Cellular senescence in aging and age-related disease: from mechanisms to therapy. Nature Medicine. 2015;**21**(12):1424, 35

[9] Al-Bari AA, Al Mamun A. Current advances in regulation of bone homeostasis. FASEB bioAdvances. 2020;**2**(11):668, 679

[10] Baker DJ et al. Clearance of p16Ink4a-positive senescent cells delays ageing-associated disorders. Nature. 2011;**479**(7372):232, 6

[11] Riggs BL, Khosla S, Melton LJ 3rd. Sex steroids and the construction and conservation of the adult skeleton. Endocrine Reviews. 2002;**23**(3):279, 302

[12] Drake MT, Clarke BL, Lewiecki EM. The pathophysiology and treatment of osteoporosis. Clinical Therapeutics. 2015;**37**(8):1837, 50

[13] Boyle WJ, Simonet WS, Lacey DL. Osteoclast differentiation and activation. Nature. 2003;**423**(6937):337, 42

[14] Bellido T. Osteocyte-driven bone remodeling. Calcified Tissue International. 2014;**94**(1):25, 34

[15] Parfitt AM. The coupling of bone formation to bone resorption: a critical analysis of the concept and of its relevance to the pathogenesis of osteoporosis. Metabolic Bone Disease & Related Research. 1982;**4**(1):1, 6

[16] Sims NA, Martin TJ. Coupling signals between the osteoclast and osteoblast: How are messages transmitted between these temporary visitors to the bone surface? Front Endocrinol (Lausanne). 2015;**6**:41

[17] Qiu S et al. Relationships between osteocyte density and bone formation rate in human cancellous bone. Bone. 2002;**31**(6):709, 11

[18] Heaney RP. Is the paradigm shifting? Bone. 2003;**33**(4):457, 65

[19] Meng X et al. Hypoxia-inducible factors regulate osteoclasts in health and disease. Frontiers in Cell and Development Biology. 2021;**9**:658893

[20] Cappariello A et al. Reprint of: The great beauty of the osteoclast. Archives of Biochemistry and Biophysics. 2014;**561**:13, 21

[21] Janda CY et al. Surrogate Wnt agonists that phenocopy canonical Wnt and β-catenin signalling. Nature. 2017;**545**(7653):234, 237

[22] Baron R, Kneissel M. WNT signaling in bone homeostasis and disease: from human mutations to treatments. Nature Medicine. 2013;**19**(2):179, 92

[23] Fowler TW et al. Development of selective bispecific Wnt mimetics for bone loss and repair. Nature Communications. 2021;**12**(1):3247

[24] Redlich K, Smolen JS. Inflammatory bone loss: Pathogenesis and therapeutic intervention. Nature Reviews. Drug Discovery. 2012;**11**(3):234, 50

[25] Matsuo K, Irie N. Osteoclast-osteoblast communication. Archives of Biochemistry and Biophysics. 2008; **473**(2):201, 9

[26] Klein-Nulend J et al. Bone cell mechanosensitivity, estrogen deficiency, and osteoporosis. Journal of Biomechanics. 2015;**48**(5):855, 65

[27] Khosla S, Oursler MJ, Monroe DG. Estrogen and the skeleton. Trends in Endocrinology & Metabolism. 2012;**23**(11):576, 581

[28] Khosla S. Pathogenesis of age-related bone loss in humans. The Journals of Gerontology. Series A, Biological Sciences and Medical Sciences. 2013;**68**(10):1226, 35

[29] Muraca M, Cappariello A. The role of extracellular vesicles (EVs) in the epigenetic regulation of bone metabolism and osteoporosis. International Journal of Molecular Sciences. 2020;**21**(22):8682

[30] Pignolo RJ et al. Targeting cell senescence for the treatment of age-related bone loss. Current Osteoporosis Reports. 2019;**17**(2):70, 85

[31] Vanderschueren D et al. Sex steroid actions in male bone. Endocrine Reviews. 2014;**35**(6):906, 60

[32] Khosla S. Update on estrogens and the skeleton. The Journal of Clinical Endocrinology and Metabolism. 2010;**95**(8):3569, 77

[33] Almeida M et al. Skeletal involution by age-associated oxidative stress and its acceleration by loss of sex steroids. The Journal of Biological Chemistry. 2007;**282**(37):27285, 27297

[34] Nakashima T et al. Evidence for osteocyte regulation of bone homeostasis through RANKL expression. Nature Medicine. 2011;**17**(10):1231, 4

[35] Liu X-L, Ding J, Meng L-H. Oncogene-induced senescence: a double edged sword in cancer. Acta Pharmacologica Sinica. 2018;**39**(10):1553, 1558

[36] Schumacher B et al. The central role of DNA damage in the ageing process. Nature. 2021;**592**(7856):695, 703

[37] Günes C, Rudolph KL. The role of telomeres in stem cells and cancer. Cell. 2013;**152**(3):390, 3

[38] Lundblad V, Szostak JW. A mutant with a defect in telomere elongation leads to senescence in yeast. Cell. 1989;**57**(4): 633, 43

[39] Maciejowski J, de Lange T. Telomeres in cancer: Tumour suppression and genome instability. Nature Reviews. Molecular Cell Biology. 2017;**18**(3): 175, 186

[40] Levy MZ et al. Telomere end-replication problem and cell aging. Journal of Molecular Biology. 1992;**225**(4):951, 60

[41] Paluvai H, Di Giorgio E, Brancolini C. The histone code of senescence. Cell. 2020;**9**:2

[42] Smith EM, Pendlebury DF, Nandakumar J. Structural biology of telomeres and telomerase. Cellular and Molecular Life Sciences. 2020;**77**(1): 61, 79

[43] Fumagalli M et al. Telomeric DNA damage is irreparable and causes persistent DNA-damage-response activation. Nature Cell Biology. 2012;**14**(4):355, 65

[44] d'Adda di Fagagna F et al. A DNA damage checkpoint response in telomere-initiated senescence. Nature. 2003;**426** (6963):194, 8

[45] Kotsantis P, Petermann E, Boulton SJ. Mechanisms of oncogene-induced replication stress: Jigsaw falling into place. Cancer Discovery. 2018;**8**(5): 537, 555

[46] Wang L et al. Osteoblast autophagy in glucocorticoid-induced osteoporosis. Journal of Cellular Physiology. 2019;**234**(4):3207, 3215

[47] Chotiyarnwong P, McCloskey EV. Pathogenesis of glucocorticoid-induced osteoporosis and options for treatment. Nature Reviews. Endocrinology. 2020;**16**(8):437, 447

[48] Leclerc N et al. Gene expression profiling of glucocorticoid-inhibited osteoblasts. Journal of Molecular Endocrinology. 2004;**33**(1):175, 93

[49] Hardy R, Cooper MS. Bone loss in inflammatory disorders. The Journal of Endocrinology. 2009;**201**(3):309, 20

[50] Severinsen MCK, Pedersen BK. Muscle-organ crosstalk: The emerging roles of myokines. Endocrine Reviews. 2020;**41**(4):594, 609

[51] Carina V et al. Bone's response to mechanical loading in aging and osteoporosis: Molecular mechanisms. Calcified Tissue International. 2020;**107**(4):301, 318

[52] Kanis JA et al. European guidance for the diagnosis and management of osteoporosis in postmenopausal women. Osteoporosis International. 2019;**30**(1):3, 44

[53] Group, D. Patient level pooled analysis of 68 500 patients from seven major vitamin D fracture trials in US and Europe. BMJ. 2010;**340**:b5463

[54] Langdahl BL, Harsløf T. Medical treatment of osteoporotic vertebral fractures. Ther Adv Musculoskelet Dis. 2011;**3**(1):17, 29

[55] Bischoff-Ferrari HA et al. Fall prevention with supplemental and active forms of vitamin D: A meta-analysis of randomised controlled trials. BMJ. 2009;**339**:b3692

[56] Noh JY, Yang Y, Jung H. Molecular mechanisms and emerging therapeutics for osteoporosis. International Journal of Molecular Sciences. 2020;**21**(20): 7628-7650

[57] Nieves JW. Chapter 71 - Nutrients beyond calcium and vitamin D to treat osteoporosis. In: Dempster DW et al., editors. Marcus and Feldman's

Osteoporosis. Fifth ed. London, United Kingdom: Academic Press; 2021. p. 1679, 1693

[58] Vorland CJ et al. Effects of Excessive Dietary Phosphorus Intake on Bone Health. Current Osteoporosis Reports. 2017;**15**(5):473, 482

[59] Ferrari S, Cosman F. Chapter 82 - Long-term treatment strategies and goal-directed therapy. In: Dempster DW et al., editors. Marcus and Feldman's Osteoporosis. Fifth ed. London, United Kingdom: Academic Press; 2021. p. 1867, 1872

[60] Drake MT, Clarke BL, Khosla S. Bisphosphonates: Mechanism of action and role in clinical practice. Mayo Clinic Proceedings. 2008;**83**(9):1032, 45

[61] Kostenuik PJ et al. Denosumab, a fully human monoclonal antibody to RANKL, inhibits bone resorption and increases BMD in knock-in mice that express chimeric (murine/human) RANKL. Journal of Bone and Mineral Research. 2009;**24**(2):182, 95

[62] Murad MH et al. Clinical review. Comparative effectiveness of drug treatments to prevent fragility fractures: a systematic review and network meta-analysis. The Journal of Clinical Endocrinology and Metabolism. 2012;**97**(6):1871, 80

[63] Ettinger B et al. Reduction of vertebral fracture risk in postmenopausal women with osteoporosis treated with raloxifene: Results from a 3-year randomized clinical trial. Multiple Outcomes of Raloxifene Evaluation (MORE) Investigators. JAMA. 1999;**282**(7):637, 45

[64] An KC. Selective estrogen receptor modulators. Asian Spine J. 2016;**10**(4): 787, 91

[65] Munoz-Torres M, Alonso G, Raya MP. Calcitonin therapy in osteoporosis. Treatments in Endocrinology. 2004;**3**(2):117, 32

[66] Dai R et al. Cathepsin K: The action in and beyond bone. Frontiers in Cell and Development Biology. 2020;**8**:433

[67] Saag KG et al. Teriparatide or alendronate in glucocorticoid-induced osteoporosis. The New England Journal of Medicine. 2007;**357**(20):2028, 39

[68] Cui C et al. Parathyroid hormone ameliorates temporomandibular joint osteoarthritic-like changes related to age. Cell Proliferation. 2020;**53**(4):e12755

[69] Kraenzlin ME, Meier C. Parathyroid hormone analogues in the treatment of osteoporosis. Nature Reviews. Endocrinology. 2011;**7**(11):647, 56

[70] Lindsay R et al. teriparatide for osteoporosis: Importance of the full course. Osteoporosis International. 2016;**27**(8):2395, 410

[71] Diez-Perez A et al. Effects of teriparatide on hip and upper limb fractures in patients with osteoporosis: A systematic review and meta-analysis. Bone. 2019;**120**:1, 8

[72] Genant HK et al. Effects of romosozumab compared with teriparatide on bone density and mass at the spine and hip in postmenopausal women with low bone mass. Journal of Bone and Mineral Research. 2017; **32**(1):181, 187

[73] McClung MR et al. Romosozumab in postmenopausal women with low bone mineral density. The New England Journal of Medicine. 2014;**370**(5):412, 20

[74] Meunier PJ et al. The effects of strontium ranelate on the risk of

vertebral fracture in women with postmenopausal osteoporosis. The New England Journal of Medicine. 2004; **350**(5):459, 68

[75] Mi B et al. Strontium-loaded titania nanotube arrays repress osteoclast differentiation through multiple signalling pathways: In vitro and in vivo studies. Scientific Reports. 2017;**7**(1): 2328

[76] Fogelman I, Blake GM. Strontium ranelate for the treatment of osteoporosis. BMJ. 2005;**330**(7505): 1400, 1

[77] Leder BZ et al. Response to therapy with teriparatide, denosumab, or both in postmenopausal women in the DATA (Denosumab and Teriparatide Administration) study randomized controlled trial. Journal of Clinical Densitometry. 2016;**19**(3):346, 51

[78] Farr JN et al. Identification of senescent cells in the bone microenvironment. Journal of Bone and Mineral Research. 2016;**31**(11): 1920, 1929

[79] Farr JN et al. Targeting cellular senescence prevents age-related bone loss in mice. Nature Medicine. 2017;**23**(9): 1072, 1079

[80] Zhu Y et al. The Achilles' heel of senescent cells: From transcriptome to senolytic drugs. Aging Cell. 2015;**14**(4): 644, 58

[81] Yousefzadeh MJ et al. Fisetin is a senotherapeutic that extends health and lifespan. eBioMedicine. 2018;**36**:18, 28

Chapter 5

Aging and Neuropsychiatric Disease: A General Overview of Prevalence and Trends

Jelena Milić

Abstract

The increasing trend of life-expectancy is becoming a significant demographic, societal and economic challenge. Currently, global number of people above sixty years of age is 900 million, while United Nations expect this number to rise to over 1.4 billion in 2030 and over 2.5 billion by 2050. Concordant to this trend, numerous physiological changes are associated with aging and brain-related ones are associated with neuropsychiatric diseases. The main goal of this chapter is to identify the most important neuropsychiatric diseases to assess in older patients to help to promote health and prevent diseases and complications associated with chronic illness, as these changes are progressive and require important psychological and setting-related social adjustments. Findings identify several health-aspects highly present in elderly: stroke, white matter lesions, dementia rise with age, changes in levels of neurotransmitters and hormones, depression as well as the bereavement following loss of the loved one, and the most common neurodegenerative disease—Alzheimer's disease and Parkinson's. In conclusion, studying the aging process should include all developmental, circumstantial, and individual aspects of aging. This offers opportunities to improve the health of elderly by using a wide range of skills and knowledge. Thus, further studies are necessary to elucidate what can be done do to improve the aging process and health of elderly in the future.

Keywords: aging, neuropsychiatric diseases, dementia, Alzheimer's, Parkinson's, depression, bereavement

1. Introduction

Aging of the population is a global phenomenon [1], it is accelerating and becoming a significant demographic, societal and economic challenge [2]. Aging is the process of becoming older. Many societies, especially in Western countries, have already attained an older population structure than has ever been seen in the past, and a trend toward this structure is being observed also in low, and middle-income countries [3]. A population where the proportion of older people is increasing

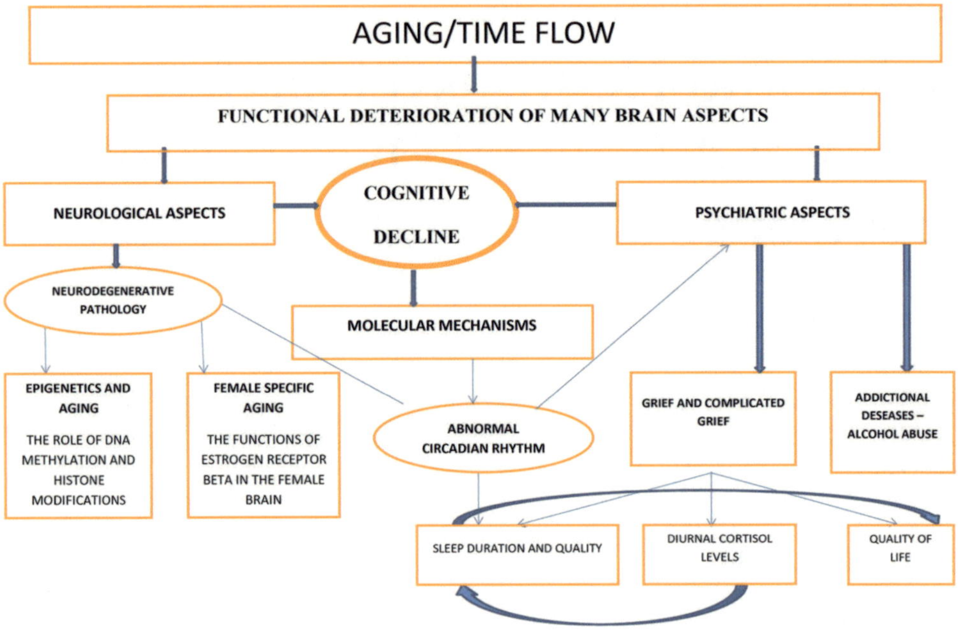

Figure 1.
The aging population and most common accompanying neuropsychiatric aspects.

is referred to as a cohort of the older adults or an aging population. The size and proportion of the global population as relates to age over time is also known as 'demographic aging' and 'population aging'. The old-age dependency ratio (OADR) is the standard indicator of population aging used in order to quantify the aging population. It expresses the amount of people aged 65 and over per 100 persons aged 15 to 64. In this way it provides an easy measure of the part of the population that have reached the traditional age of pension in relation to those of working age. In humans, aging represents the accumulation of changes in a human being over time and can encompass physical, psychological, and social changes. There are seven stages of human development. These stages are: infancy, early childhood, middle childhood, adolescence, early adulthood, middle age and old age. Aging is a gradual, continuous process of natural change that begins in early adulthood. During early middle age, many bodily functions begin to gradually decline. Although women live longer on average, they do tend to age slightly faster than men. Women also tend to make more use of cosmetic procedures in an attempt to keep the more visible effects of aging at bay. The actual rate of aging varies by person, depending on genetics, lifestyle choices and environmental factors. As the population ages, countries are slowly raising the age of pension. Therefore, it becomes increasingly relevant that the aging population is healthy. Gerontology studies all aspects of aging. Not only the physical changes, but also the mental, social and societal implications of growing older. This multidisciplinary approach offers a lot of opportunities to improve the health of older adults, using a wide range of skills and knowledge. In this chapter we present the insights in the most common gerontological neuropsychiatric aspects that appear with the ongoing aging process: neurodegenerative process—cognitive decline and epigenetics background, specific female aspect of aging, bereavement and depression, declining sleep quality and effect on quality of life via (ab)normal circadian rhythm (**Figure 1**).

2. Definition of old age and future perspectives in demographics

There is no specific age at which people are considered old or elderly. Traditionally however, age 65 has been designated as the beginning of old age. There are more than 900 million people worldwide aged 60 and over and this number is projected to grow. The United Nations expect this group to number to be 1.4 billion in 2030 and 2.5 billion by 2050 [4]. Within this group, the subcategory of "very old age people", aged 80 and over, makes up a sizable portion [5]. Societal aging can affect economic growth, patterns of work and retirement, the way that families function, the ability of governments and communities to provide adequate resources for older adults, and the prevalence of chronic disease and disability. Far-reaching economic and social adjustments will be required in most countries, as well as understanding the impact of this transformation on health and everyday living [6, 7]. Thus, it is noticeable that the aging process represent a social problem. Some of the additional aspects attached to this are: living alone, family violence, loss of a spouse, aging-related cognitive impairments and disabilities, and transport issues. Consequences of social isolation can be dangerous, particularly for individuals already predisposed to health problems. Aging is a universal trait that is observed across the evolutionary spectrum. Determining the causal underlying cellular and molecular processes that deteriorate with age and lead to increased disease susceptibility and frailty is critical if we are to meet the growing healthcare needs of an aging human populations.

3. Overview on some of the aging theories and classifications

A number of theories about aging exist: According to Rate-of-living theory the faster an organism's metabolism is, the shorter its lifespan will be. Wear and tear theory states, that our bodies (tissue and cells) simply get damaged and wear out by usage. Cross-linking theory postulates, that over time proteins form cross-links that accumulate and slow down the processes in the body and damage cells. Aging is a step-wise process and can be presented in 5 stages of aging: Stage 1—Independence, Stage 2—Interdependence, Stage 3—dependency, Stage 4—Crisis Management, and Stage 5—End of Life. Further, there are 2 different types of aging. Intrinsic aging occurs naturally as we grow older and is largely a product of heredity. Extrinsic aging is based almost entirely on external factors. We can also emphasize that he main characteristics of aging or the most visible one is presented on skin. Skin changes are among the most visible signs of aging. Evidence of increasing age includes wrinkles and sagging skin. However more important changes are connected to cognitive and overall psychological functioning.

4. Aging population and neuropsychiatric diseases

There are a number of physiological changes associated with aging into the elderly years (i.e. 65 and over). These changes are progressive. Because of, and alongside with, these changes come important psychological and social adjustments. Due to the brain morphological changes that come with aging, many elderly people are affected by neuropsychiatric diseases [8, 9]. The shrinking of the overall volume of the brain, a process which starts in humans when they're in their 30's, increases around the age of 60. Some areas of the brain, however, shrink more, and faster, than others. The prefrontal cortex, at the front of the frontal lobe, and the hippocampus, in the limbic system, in particular shrink with age; areas of the brain associated with complex mental activities like learning, planning and Cole et al. [8] memorizing. Moreover, incidence of stroke, white matter lesions, and dementia rise with age, as does level of memory impairment and there are changes in levels of neurotransmitters and hormones. Some amount of brain shrinkage occurs naturally as people age. Other potential causes of brain shrinkage include injury, certain diseases and disorders, infections, and alcohol use. One might wonder what the difference is between normal brain changes with aging and pathological changes with brain diseases. A moderate decline in some cognitive abilities is expected and part of healthy aging. In the case of severe and progressive decline of the cognitive functions is considered dementia. Dementia affects almost all aspects of a patient's life negatively. In **Figure 2** we present the warning signs preceding the diagnosis as well as classification of Dementia.

Even though dementia is the first association many people have in regard to aging, it is in fact depression that is the most common psychiatric disorder, affecting up to 50% of elderly people. The significance of depression is very important to understand. Depression, a type of mood disorder that is, as mentioned, the most prevalent mental health problem among older adults, is associated with distress and suffering. It also can lead to impairments in physical, mental, and social functioning. Further, Alzheimer's Disease (AD) and Parkinson's disease (PD) are the most common neurodegenerative disease [4, 10]. Worldwide, the global prevalence of dementia is estimated to be 3.9 % in people aged 60+ years, most suffering from AD [11]. Dementia is commonly accompanied by several neuropsychiatric symptoms, like agitation, apathy, delusions, depression, hallucinations and sleep impairment. These symptoms can in some cases cluster into syndromes. Operational criteria are proposed for specific psychotic and mood disturbances associated with dementia which make dementia an even bigger problem for both patients and caregivers. Alzheimer's disease and related forms of dementia have

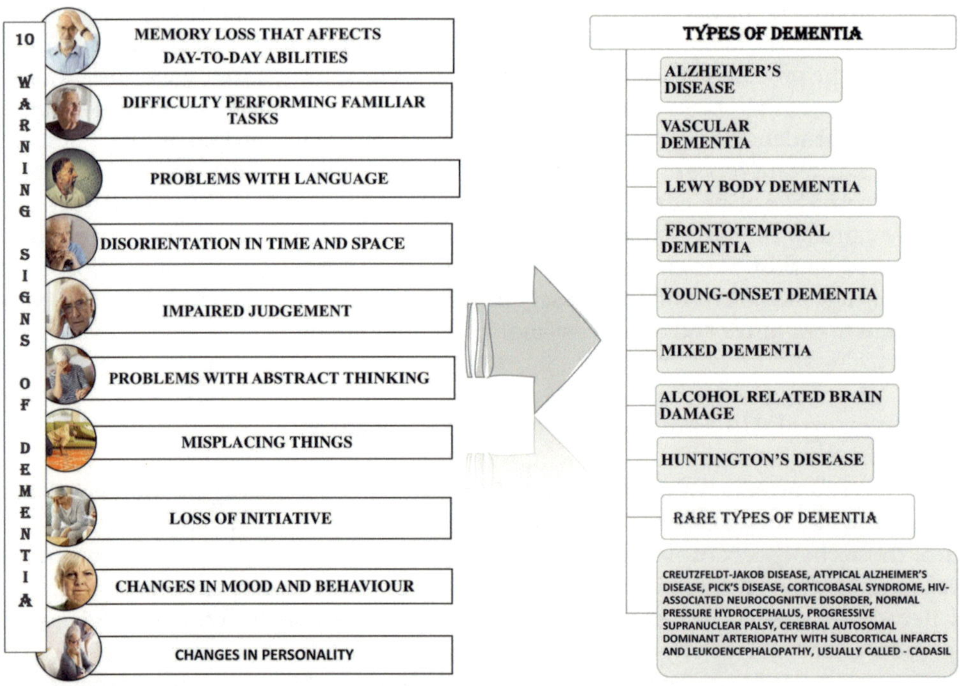

Figure 2.
Dementia: warning signs and classification.

neuropsychiatric symptoms as a core feature. Previously, these symptoms were considered part of the late stages of the disease. Currently, these symptoms, like mild cognitive impairment, are known to be present in very early phases of the disease. It is sometimes not easy to tell the difference between dementia and normal aging. However, dementia is a progressive disease that causes cognitive function to break down abnormally, causing cognitive and physical symptoms that worsen over time. On the other hand, normal aging is much more minor, with changes resulting from a natural slowing or decrease in efficiency in the body. Due to aging population, the expectation is, that the number of people suffering from dementia will double every 20 years [11]. Neuropsychiatric diseases already make up 17.4% of Years Lived with Disability (YLPs) and 6.6% of all Disability Adjusted Life Years (DALYs), placing a high economic burden on national economies. These numbers are projected to increase in the future [12]. Moreover, elderly people, especially older women, are particularly vulnerable to the adverse effects of alcohol and, alcohol use disorders in this subgroup, are often overlooked or misdiagnosed [13]. It is expected, that in the coming years, the absolute number of elderlies with problems related to alcohol will rise, further increasing the number of cognitive and physiologically impaired elderly [14–16]. Each of these substantial neuropsychiatric morbidities are often accompanied with behavioral manifestations.

5. How does epigenetics affect aging?

Aging is an inevitable outcome of life, characterized by progressive decline in tissue and organ function and increased risk of mortality. Accumulating evidence links aging to genetic and epigenetic alterations. The end result of epigenetic changes during aging is altered local accessibility to the genetic material, leading to aberrant gene expression, reactivation of transposable elements, and genomic instability. Several lifestyle factors have been identified that might modify epigenetic patterns, such as diet, obesity, physical activity, tobacco smoking, alcohol consumption, environmental pollutants, psychological stress, and working on night shifts. Epigenetics is an emerging factor in development of neurodegenerative diseases. The last decade broadened our knowledge about the etiology of AD and PD [17–20]. It is now widely accepted, that a strong genetic component contributes to the development of AD and PD, with chromosome aberrations and gene mutations playing an important role in these neurodegenerative disorders [20, 21]. One of the factors that facilitated understanding of genetic modifications is epigenetics, and its development [22]. Epigenetics studies the influence of factors external to DNA on the genes, the prefix "epi- "in this case meaning "on top of" or "above". These factors, like behavior and environment, do not change the sequence of the DNA, but alter the structure of the DNA. This influences how DNA is 'read', effectively turning genes "on" or "off". These changes can be inherited, as gene expression altered via histione proteins and DNA methylation are known to be passed on to offspring [23]. The influence of epigenetical mechanisms on the etiology of AD and PD have already been suggested by several studies [24–26]. DNA methylation and histone acetylation are also thought to play a role in depression [27], which is an important correlate of neurodegenerative diseases [28]. Despite this evidence, to date, there is not yet a comprehensive assessment on the role of epigenetic mechanisms, such as DNA methylation or histone modifications, in the development of neurological diseases. However, it is known that epigenetic reprogramming reverses most if not all of the age-related epigenetic modifications. Harnessing partial reprogramming seems the most promising therapy to treat aging

6. Aging and neuropsychiatric disorders in women: role of menopause transition and estrogen

The aging process has a dissimilar effect on the different genders. Whereas the male body changes gradually over the years, the female body changes much more abruptly during menopause. In the end, everyone has to deal individually with the effects of aging on their sexual functioning, like erectile dysfunction and vaginal dryness. When it comes to aging, women's levels of the estrogen hormone begin to decline much earlier and much more quickly than men's levels of the testosterone hormone do. This quicker decline in hormone levels is the reason that men seem to age much more slowly than women do. However, women enjoy a longer lifespan, which puts them more at risk of psychoneurological diseases [29]. Hypercoagulable states due to pregnancy and birth control pills pose another neuropsychiatric health risk unique to women [30]. The majority of diagnoses for common neurological diseases are for women, including AD, PD and depression. Among diagnosed women, these diagnoses are more prevalent in women past their menopause [31]. Besides this, a number of neurological diseases, stroke among them, are more damaging to women than to men [32]. Upon reaching menopause, because of the shift in hormonal balance, women suffer the loss of the protective anti-inflammatory benefits of estrogen. Hormone replacement therapy can't adequately counter the state of hypoestrogenia [29]. The many physiological functions of estrogen are mediated by Estrogen Receptors Alpha and Beta (ERβ). ERβ is a recent discovery, but extensively present in the brain and functional in both males and females. It is as of yet unclear, if ERβ can be a basis for new therapeutic approaches that can help treat or prevent neuropsychiatric diseases in females in the menopausal and postmenopausal stages. Data further suggest that estrogen signaling varies depending on age and stage of menopause. Growing older and going through menopause, women become more dependent on alcohol and related problems increase [33]. Because women metabolize alcohol differently from men, they are less resistant to its detrimental effects (**Figure 3**), and have the tendency to suffer earlier on from diseases and other repercussions of alcohol usage than men [33]. Also, women have more difficulty gaining access to treatment and recovering from alcohol dependence [33].

7. Aging and bereavement: impact on quality of life

Almost all persons experience the loss of a close person during the life span [34–36]. With aging, this might occur once or several different times. Older adults experience grief at a higher rate than younger adults or children. Spousal loss is common in older adults as well as the death of friends, siblings and cousins. Because of ageism, growing older comes with a measure of anxiety. Combined with the various other obstacles of aging, this can make grief a complex affair in older persons. Change and loss are integral to life, but can be very painful. As we use the terms berievement and grief interchageblz through this chapter, we would like to define that bereavement is the state of being in grief with there being different stages of bereavement. Bereavement in older age can lead to loneliness and an increased likelihood of depression, and it is appalling that older bereaved people aren't being offered the support and access to services that could make a huge difference to their well-being. The sentiments most often occurring within the period of bereavement comprise changed modality of feelings, reoccurring thoughts about the loss, somatic symptoms that sometimes lead to physical illness due to profound sadness that dominates bereavement and markable change in behavioral patterns in an attempt to escape previously

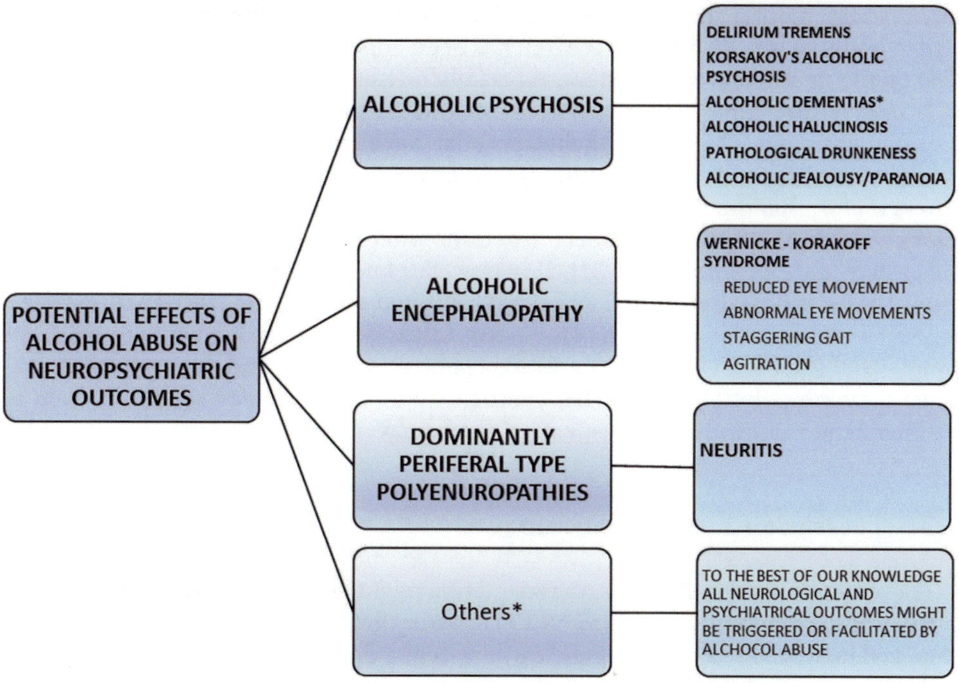

Figure 3.
*Neuropsychiatric outcomes: harmful effects of alcohol abuse. *Others: Alcoholic liver disease: women are more likely than men to develop alcoholic hepatitis (liver inflammation) and to die from cirrhosis. Brain disease: most alcoholics have some loss of mental function, reduced brain size, and changes in the function of brain cells. Research suggests that women are more vulnerable than men to alcohol-induced brain damage. Cancer: many studies report that heavy drinking increases the risk of breast cancer. Alcohol also is linked to cancers of the digestive track and of the head and neck (the risk is especially high in smokers who also drink heavily). Heart disease: chronic heavy drinking is a leading cause of cardiovascular disease. Among heavy drinkers, women are more susceptible to alcohol-related heart disease, even though women drink less alcohol over a lifetime than men.*

created behavioral routines in connection with the loved one or another loss-related entity or circumstances. Bereavement is accompanied by several well recognized feelings. The most noticed feelings in the bereavement period are: shock, desensitization, sorrow, refutation, misery, yearning annoyance, guilt, isolation, melancholy and incapability. Notwithstanding the fact that loss is an upsetting experience, the majority of people who go through it manage to recuperate within the time window of six months to a year [37]. By estimation 10 to 20% of persons suffering bereavement remain captive to remembrance, sorrow, and guilt, slipping into a long period of grief [38, 39]. This is a complicated condition named Prolonged Grief Disorder (PGD) or 'complicated grief' [38, 39]. Grief is separate from depression or anxiety. It can seriously hinder day-to-day social interactions and interpersonal relations, negatively influencing everyday life [40]. Evidence shows, that both acute and complicated grief can cause changes in the daily cycle [41], altereds mental state [42] and deviations in patterns of nourishment [42] and sleep [43]. Considering the amount of change that can be affected on everyday activities, bereavement can potentially greatly diminish quality of life. Losing a loved one is a very unsettling occurrence, that can set off powerful emotional stress. Under this newly created condition, the hypothalamic-pituitary-adrenocortical (HPA) axis is stimulated and activates the secretion of cortisol into the bloodstream stressor adaptation. Also, cortisol dysregulation triggers the problematic alcohol use leading to dependency [44]. These further effects memory loss, mental and physical disability, lower quality of life and increased chance of death [45, 46]. Little research has been done, however, on the influence of PGD on the neuroendocrine system, especially on the pattern of cortisol secretion [47]. The relationship between bereavement and disturbed sleeping patterns has already mentioned [41, 42, 48] . Schwartz et al. have studied the effect of changes in general health on the quality of life (QoL) of people suffering bereavement [49]. Others found, that the degree in which a person managed to maintain a degree of vigor and mental, physical and social reactions while grieving, influenced his sense of self-esteem and fulfillment [49–51]. However, our knowledge regarding the associations of grief and health-related outcomes is limited on a studies with small sample size and a cross-sectional design [41, 52–56]. Likewise, association between grief and complicated grief with QoL and its domains remains unclear and the available evidence examining this hypothesis has yet to be rigorously reviewed in order to help us to understand most influential aspects of the bereavement on the everyday life style.

8. Grief cessation and its determinants

For some bereaved individuals, the adaptation might be complicated, slowed, or halted, leading to incessant grief [57, 58]. This lasting grief impairs daily functioning, sleep, and increased risk of cancer and cardiovascular disease [59, 60]. The severity of grief relates to severity of impairment. Grief can change personality of a person who suffers, both on a temporary or more permanent basis based on various factors including how profound the loss was, internal coping skills, support system, general temperament, stress tolerance, and outlook on life. Therefore, detecting the determinants and predictors of traumatic grief is of crucial importance for identifying bereaved individuals in greatest risk for long-term dysfunction, and guide the development of novel interventions for the disorder [61]. Several factors have been suggested to influence the duration and severity of grief; gender plays an important role in bereavement. After losing a partner, the negative effect on health and general quality of life is gretaer with men than with women [62]. Members of both sexes show higher mortality rates when bereaved compared to non-bereaved individuals; this relative increase, however, is higher in men suffering bereavement [62]. Prior study suggests this is because men have a weaker support network and worse abilities for coping and self-empowerment than women [63]. Additionally, age is an important factor determining the length of the bereavement period. According to Strobe and Schut, bereaved persons that are younger, encounter more complications after a loss. Namely, including more serious health consequences, grief symptoms, and psychological and physical symptoms [62]. It is probable, that the reason for this difference is that younger grievers experiencing more unexpected and sudden losses, which causes more severe grief. It might be that older grievers develop better coping strategies due to life experience. Moreover, in elderly, emotions tend to be damped down, and it's less common for people to respond excitable to worries [64]. Even though younger grievers may experience more complications in acute phase of grieving, previous studies suggest that they recover more quickly. This might be supported by more access to various types of social support [62]. In conjunction with mentioned findings, series of past studies aimed to understand the grief and determinants and predictors; findings showed that grief was associated with functional impairment, gender [62], coping strategies [65], ethnicity [66], employment [60], spousal bereavement, sleep disturbances [41, 42, 48] , high-risk behaviors and increased risk of cancer and cardiovascular disease [59, 60]. Previously published evidence indicated that there is a high chance of development of commodities and psychiatric models such as: intensive stress, clinically relevant depressive symptoms, episodes of major depression and anxiety related disorders [38]. There are also chances of onset of decreased function of the immune system [67] significantly modified quality of life that decreases in accordance to severity of bereavement [68], as well as suicide tendencies and events and risk of elevated overall mortality incidence [69]. To the best of our knowledge, only few longitudinal studies on this topic had been performed so far, with a short follow-up (up to 24 months). None of the previous studies explored explore the determinants of grief bereavement bereavement related to long-term bereavement and different kind of loss in a normal population cohort with sufficient power [9, 39, 67, 70]. Further studies are necessary in order to make a stronger conclusion.

9. Why is it important to understand the elderly and the aging?

The study of elderly and the aging helps us understand the society in which we live, and it also alerts us to certain processes and problems that we may experience as we grow into old age. The fact that we are all facing the longer life span make us aware that is necessary to plan better for the course of events expected on this prolonged journey. Longer lives must be planned for. Societal aging may affect economic growth and many other issues, including the sustainability of families, the ability of states and communities to provide resources for older citizens, and international relations. Older people are important members of any society and therefore have the right to live in dignity in later life. It includes psychological and health aspects, continuing to participate in society, and ensuring a safe source of income. This also needs to be very well planned for on behalf of states and communities.

10. Cultural distinctions facing elderly in different societies

Many researchers have tried to identify different potential clarification for cross-cultural differences in overview of aging. Previously published evidence focused on socioeconomic predictors has reveale that higher levels of economic development and industrialization are associated with ominous attitudes towards aging and a lower societal status of elderly [71, 72]. Modernization theory, that is commonly in use when elucidating the activity of modernization within different societies. The theory identifies the inner aspects of a country that has been modernized while in the same time hypothesizing that with adequate help and methodology the rest of the countries can be changed in the same way and become more developed. Previous studies suggested [73, 74] that a shift towards industrialized models of mass-production erode the societal status of older adults, depreciate their insights that derived from reach experience, deconstruct the ties between the nuclear and procreational families through urbanization, and refocus the power over the means of production from elderly members of the families to industrial entities [73]. Although intuitively appealing, modernization theory has been criticized as an oversimplification [75]. Eastern/Asian compared to Western cultures is the model that has been most explored and discussed when it comes to detecting of the influence of cultural values and beliefs on aging behavioral pathways (see [76] for a review). The main point of this model comes from the idea that Asian societies are based on Confucian values of filial piety and the practice of ancestor worship which are thought to promote positive views of aging and high esteem for older adults ([77] for a review). On the other hand, the Western societies, were believed to be oriented towards youngsters and to hold not so positive views about the aging process and the older adults [72]. However, verifiable evidence for the proposed East-West differences is hard to come by. In summary, there is some evidence that both socioeconomic development and cultural values and beliefs may matter for cross-cultural differences in aging attitudes. So far, we know very limited arguments about whether the course of life and the growth —and particularly development in elderly years of life are globally a cultural similar phenomenon or culture-limited. Although the findings are limited in several important aspects one of them is sufficiently explored and that is how the country perceives neuropsychiatric aspects of aging. There are differences in neuropsychiatric aging between different cultures, both in the speed of decline of different functions and the perception of the decline itself [78]. Most research in this area focuses on the USA and parts of Asia, leaving large parts of the globe still underrepresented. Instruments developed in Western countries also lead to misrepresentation or over- or under-diagnosis of mental aging in other settings. A certain cultural awareness thus needs to be present to accurately treat people from different cultures. On the other hand, recognizing the existence of cultural differences also lead to a search for more objective methods that can be used equally for all cultures to determine neuropsychiatric decline. New instruments are being developed to address cross-cultural neuropsychiatry of aging [79], which is especially important for globalized, multi-ethnic societies [80–90].

11. Conclusion

In drawing to a close, this general overview of prevalence and trends of neuropsychiatric disease in aging population indicates that this global phenomenon of aging and increasing neuropsychiatric diseases in aged population is accelerating. The aging of the population is dominantly guided by declines in reproductive potentials of humans resulting in less pregnancies and improvements in healthcare and consequent longevity. This increase in overall life expectancy and the consequent number of elderly people in overall global population presents a significant demographic, societal and economic challenge. Therefore, the present moment is the right time to act towards more research in this area as well as the area of aging in general. The main focus of further aging research should be on the physiological changes associated with aging in the elderly years, since these changes are progressive and they trigger the important psychological and social adjustments. The leading research focus has been for some time given to the brain morphological changes that come with aging, as many elderly people are affected by neuropsychiatric diseases such as stroke, white matter lesions, and dementia, as well as overall changes in levels of neurotransmitters and hormones that are influencing the decrease in overall health and quality of life. As already emphasized, depression is the most common psychiatric disorder, affecting up to 50% of elderly people, and therefore also a highly important research topic in the aging research field. Depression is in old age often accompanied by bereavement that follows the loss of the loved one. As the loss of the loved one (dominantly a partner or sibling) is very common in elderly population. Even though, the relationship between depression and bereavement is complicated and often understudied and misleading to even the most experienced clinicians, the relationships between symptoms and treatment of bereavement and depression were disentangled in Diagnostic and Statistical Manual of Mental Disorders, the 3rd edition, a comprehensive classification of officially recognized psychiatric disorders, published by the American Psychiatric Association, where the mental health professionals have had introduced the exclusion of bereavement as the prerogative for diagnosis of major depressive disorder. This was introduced in order to protect against bereavement being mistaken for depression, mislabeled as an illness, pathologized, and/or inappropriately treated. However, both depression and bereavement are two psychiatric states that we need to keep a close focus on due to increasing incidence. Furthermore, Alzheimer's Disease and Parkinson's disease are the most common neurodegenerative disease in the elderly. Distinctively, women's levels of the estrogen hormone begin to decline much earlier and much more quickly than men's levels of the testosterone hormone, and therefore men seem to age much more slowly than women do. However, women enjoy a longer lifespan, which puts them more at risk of psychoneurological diseases. Therefore, when it comes to women the focus of the future aging research should comprise: hypercoagulable states due to pregnancy, Alzheimer's Disease, Parkinson's disease and depression, as there are present in both genders. As previously mentioned, among diagnosed women, these diagnoses are more prevalent in women past their menopause [31]. Next, a number of neurological diseases, stroke being one of them, are more damaging to women than to men [32] and therefore future research should focus on exploring the risk factors and potential treatments. Also, due to the fact that women metabolize alcohol differently from men, they are less resistant and have the tendency to suffer earlier on from diseases and other repercussions of alcohol usage than men [33], the studies that are tailor-targeted for diagnostics and treatments of addiction diseases for female gender should be conducted more.

To conclude, the multidisciplinary approach should be considered when studying the aging process. This should include all aspects of aging, by exploring not only the physical changes, but also the mental, social and societal implications of growing older offers a lot of opportunities to improve the health of older adults, using a wide range of skills and knowledge, and future studies and overall research focus should be following the gerontological updates and broaden it to optimal extent.

Author details

Jelena Milić
Department for Methodological Principles and Standards of Integrated Health Information System and Reporting, Institute of Public Health of Serbia "Dr. Milan Jovanovic Batut", Belgrade, Serbia

*Address all correspondence to: dr.jelena.milic@gmail.com

References

[1] Akiyama H. Aging well: An update. Nutrition Reviews. 2020;**78**(12 Suppl):3-9

[2] Oeppen J, Vaupel JW. Demography. Broken limits to life expectancy. Science. 2002;**296**(5570):1029-1031

[3] Salomon JA, Wang H, Freeman MK, Vos T, Flaxman AD, Lopez AD, et al. Healthy life expectancy for 187 countries, 1990-2010: A systematic analysis for the Global Burden Disease Study 2010. Lancet. 2012;**380**(9859):2144-2162

[4] United Nations Population Fund (UNFPA). Ageing in the Twenty-First Century: A Celebration and a Challenge. 2015. Available from https://www.unfpa.org/sites/default/files/pub-pdf/Ageing%20report.pdf

[5] Forman DE, Berman AD, McCabe CH, Baim DS, Wei JY. PTCA in the elderly: The "young-old" versus the "old-old". Journal of the American Geriatrics Society. 1992;**40**(1):19-22

[6] Boz C, Ozsarı SH. The causes of aging and relationship between aging and health expenditure: An econometric causality analysis for Turkey. The International Journal of Health Planning and Management. 2020;**35**(1):162-170

[7] Colchero F, Rau R, Jones OR, Barthold JA, Conde DA, Lenart A, et al. The emergence of longevous populations. Proceedings of the National Academy of Sciences of the United States of America. 2016;**113**(48):E7681-E7690

[8] Cole JH, Marioni RE, Harris S, Deary EJ. Brain age and other bodily 'ages': Implications for neuropsychiatry. Molecular Psychiatry. 2019;**24**:266-281

[9] Taneri PE, Kiefte-de Jong JC, Bramer WM, Daan NM, Franco OH, Muka T. Association of alcohol consumption with the onset of natural menopause: A systematic review and meta-analysis. Human Reproduction Update. 2016;**22**(4):516-528

[10] Djernes JK. Prevalence and predictors of depression in populations of elderly: A review. Acta Psychiatrica Scandinavica. 2006;**113**(5):372-387

[11] Qiu C, Kivipelto M, von Strauss E. Epidemiology of Alzheimer's disease: Occurrence, determinants, and strategies toward intervention. Dialogues in Clinical Neuroscience. 2009;**11**(2):111-128

[12] WHO. Mental health and older adults. 2016. Available from: http://www.who.int/mediacentre/factsheets/fs381/en/

[13] Bratberg GH, Wilsnack R, Havas Haugland S, Krokstad S, Sund ER, Bjorngaard JH. Gender differences and gender convergence in alcohol use over the past three decades (1984-2008), The HUNT Study, Norway. BMC Public Health. 2016;**16**:723

[14] Briggs W, Magnus V, Lassiter P, Patterson A, Smith L. Substance use, misuse, and abuse among older adults: Implications for clinical mental health counselors. JMHC. 2011;**33**(2):112-127

[15] Connell H, Chin A-V, Cunningham C, Lawlor B. Alcohol use disorders in elderly people–redefining an age old problem in old age. BMJ. 2003;**327**(7416):664

[16] Han B, Gfroerer JC, Colliver JD, Penne MA. Substance use disorder among older adults in the United States in 2020. Addiction. 2009;**104**(1):88-96

[17] Bertram L, McQueen MB, Mullin K, Blacker D, Tanzi RE. Systematic

meta-analyses of Alzheimer disease genetic association studies: The AlzGene database. Nature Genetics. 2007;**39**(1):17-23

[18] Farrer MJ. Genetics of Parkinson disease: Paradigm shifts and future prospects. Nature Reviews: Genetics. 2006;7(4):306-318

[19] Gusella JF, MacDonald ME, Ambrose CM, Duyao MP. Molecular genetics of Huntington's disease. Archives of Neurology. 1993;**50**(11):1157-1163

[20] Karch CM, Goate AM. Alzheimer's disease risk genes and mechanisms of disease pathogenesis. Biological Psychiatry. 2015;77(1):43-51

[21] Ghani M, Lang AE, Zinman L, Nacmias B, Sorbi S, Bessi V, et al. Mutation analysis of patients with neurodegenerative disorders using NeuroX array. Neurobiology of Aging. 2015;**36**(1):e549

[22] Jakovcevski M, Akbarian S. Epigenetic mechanisms in neurological disease. Nature Medicine. 2012;**18**(8):1194-1204

[23] Henikoff S, Matzke MA. Exploring and explaining epigenetic effects. Trends in Genetics. 1997;**13**(8):293-295

[24] Bollati V, Galimberti D, Pergoli L, Dalla Valle E, Barretta F, Cortini F, et al. DNA methylation in repetitive elements and Alzheimer disease. Brain, Behavior, and Immunity. 2011;**25**(6):1078-1083

[25] Mogi M, Harada M, Narabayashi H, Inagaki H, Minami M, Nagatsu T. Interleukin (IL)-1 beta, IL-2, IL-4, IL-6 and transforming growth factor-alpha levels are elevated in ventricular cerebrospinal fluid in juvenile parkinsonism and Parkinson's disease. Neuroscience Letters. 1996;**211**(1):13-16

[26] Urdinguio RG, Sanchez-Mut JV, Esteller M. Epigenetic mechanisms in neurological diseases: Genes, syndromes, and therapies. Lancet Neurology. 2009;**8**(11):1056-1072

[27] Sun H, Kennedy PJ, Nestler EJ. Epigenetics of the depressed brain: Role of histone acetylation and methylation. Neuropsychopharmacology. 2013;**38**(1):124-137

[28] Rickards H. Depression in neurological disorders: Parkinson's disease, multiple sclerosis, and stroke. Journal of Neurology, Neurosurgery, and Psychiatry. 2005;**76**(Suppl 1):i48-i52

[29] O'Neal MA. Neurologic diseases in women: Five new things. Neurology Clinical Practice. 2013;**3**(3):217-223

[30] Katz D, Beilin Y. Disorders of coagulation in pregnancy. BJA. 2015;**115**(suppl_2):ii75-ii88

[31] Mary VS. Psychopathology in women and men: Focus on female hormones. The American Journal of Psychiatry. 1997;**154**(12):1641-1647

[32] Hofman A, de Jong PTVM, van Duijn CM, Breteler MMB. Epidemiology of neurological diseases in elderly people: What did we learn from the Rotterdam Study? Lancet Neurology. 2006;**5**(6):545-550

[33] Epstein EE, Fischer-Elber K, Al-Otaiba Z. Women, aging, and alcohol use disorders. Journal of Women & Aging. 2007;**19**(1-2):31-48

[34] Boelen PA, Hoijtink H. An item response theory analysis of a measure of complicated grief. Death Studies. 2009;**33**(2):101-129

[35] Monk TH, Begley AE, Billy BD, Fletcher ME, Germain A, Mazumdar S, et al. Sleep and circadian rhythms in spousally bereaved seniors.

Chronobiology International. 2008a;**25**(1):83-98

[36] Shear MK. Clinical practice. Complicated grief. New England Journal of Medicine. 2015;**372**(2):153-160

[37] Kacel E, Gao X, Prigerson HG. Understanding bereavement: What every oncology practitioner should know. The Journal of Supportive Oncology. 2011;**9**(5):172-180

[38] Prigerson HG, Bierhals AJ, Kasl SV, Reynolds CF, Shear MK, Day N, et al. Traumatic grief as a risk factor for mental and physical morbidity. The American Journal of Psychiatry. 1997;**154**(5):616-623

[39] Prigerson HG, Horowitz MJ, Jacobs SC, Parkes CM, Aslan M, Goodkin K, et al. Prolonged grief disorder: Psychometric validation of criteria proposed for DSM-V and ICD-11. PLoS Medicine. 2009;**6**(8):e1000121

[40] Newson RS, Boelen PA, Hek K, Hofman A, Tiemeier H. The prevalence and characteristics of complicated grief in older adults. Journal of Affective Disorders. 2011;**132**(1-2):231-238

[41] Monk TH, Germain A, Reynolds CF. Sleep disturbance in bereavement. Psychiatric Annals. 2008b;**38**(10):671-675

[42] Hall CA, Reynolds CF, Butters M, Zisook S, Simon N, Corey-Bloom J, et al. Cognitive functioning in complicated grief. Journal of Psychiatric Research. 2014;**0**:20-25

[43] Milic J et al. The longitudinal and cross-sectional associations of grief and complicated grief with sleep quality in older adults. Behavioral Sleep Medicine. 2019;**17**(1):31-40

[44] Stephens MA, Wand G. Stress and the HPA axis: Role of glucocorticoids in alcohol dependence. Alcohol Research: Current Reviews. 2012;**34**(4):468-483

[45] Kumari M, Shipley M, Stafford M, Kivimaki M. Association of diurnal patterns in salivary cortisol with all-cause and cardiovascular mortality: Findings from the Whitehall II study. The Journal of Clinical Endocrinology and Metabolism. 2011;**96**(5):1478-1485

[46] Mura G, Cossu G, Migliaccio GM, Atzori C, Nardi AE, Machado S, et al. Quality of life, cortisol blood levels and exercise in older adults: Results of a randomized controlled trial. Clinical Practice and Epidemiology in Mental Health. 2014;**10**:67-72

[47] Saavedra Perez HC, Direk N, Milic J, Ikram MA, Hofman A, Tiemeier H. The impact of complicated grief on diurnal cortisol levels two years after loss: A population-based study. Psychosomatic Medicine. 2017;**79**(4):426-433

[48] Kowalski SD, Bondmass MD. Physiological and psychological symptoms of grief in widows. Research in Nursing & Health. 2008;**31**(1):23-30

[49] Schwartz CE, Sprangers MA. Methodological approaches for assessing response shift in longitudinal health-related quality-of-life research. Social Science & Medicine. 1999;**48**(11): 1531-1548

[50] Cousson-Gelie F, de Chalvron S, Zozaya C, Lafaye A. Structural and reliability analysis of quality of relationship index in cancer patients. Journal of Psychosocial Oncology. 2013;**31**(2):153-167

[51] Ozer ZC, Firat MZ, Bektas HA. Confirmatory and exploratory factor analysis of the caregiver quality of life index-cancer with Turkish samples. Quality of Life Research. 2009;**18**(7):913-921

[52] Boelen PA, Lancee J. Sleep difficulties are correlated with emotional problems following loss and residual symptoms of effective prolonged grief disorder treatment. Depression Research and Treatment. 2013;**2013**:739-804

[53] Germain A, Caroff K, Buysse DJ, Shear MK. Sleep quality in complicated grief. Journal of Traumatic Stress. 2005;**18**(4):343-346

[54] Maytal G, Zalta AK, Thompson E, Chow CW, Perlman C, Ostacher MJ, et al. Complicated grief and impaired sleep in patients with bipolar disorder. Bipolar Disorders. 2007;**9**(8):913-917

[55] Purebl G, Pilling J, Konkoly TB, Bodizs R, Kopp M. [Are oppressive dreams indicators in bereavement?] Van-e a nyomaszto almoknak indikatorszerepuk a gyaszban? Ideggyógyászati Szemle. 2012;**65**(7-8):261-265

[56] Spira AP, Stone K, Beaudreau SA, Ancoli-Israel S, Yaffe K. Anxiety symptoms and objectively measured sleep quality in older women. The American Journal of Geriatric Psychiatry. 2009;**17**(2):136-143

[57] Boelen PA, van den Bout J. Complicated grief and uncomplicated grief are distinguishable constructs. Psychiatry Research. 2008;**157**(1-3): 311-314

[58] Shear K, Monk T, Houck P, Melhem N, Frank E, Reynolds C, et al. An attachment-based model of complicated grief including the role of avoidance. European Archives of Psychiatry and Clinical Neuroscience. 2007;**257**(8):453-461

[59] Simon NM, Pollack MH, Fischmann D, Perlman CA, Muriel AC, Moore CW, et al. Complicated grief and its correlates in patients with bipolar disorder. The Journal of Clinical Psychiatry. 2005;**66**:1105-1110

[60] Simon NM, Shear KM, Thompson EH, Zalta AK, Perlman C, Reynolds CF, et al. The prevalence and correlates of psychiatric comorbidity in individuals with complicated grief. Comprehensive Psychiatry. 2007;**48**:395-399

[61] Currier JM, Neimeyer RA, Berman JS. The effectiveness of psychotherapeutic interventions for bereaved persons: A comprehensive quantitative review. Psychological Bulletin. 2008;**134**(5):648-661

[62] Stroebe W, Schut H. Risk factors in bereavement outcome: A methodological and empirical review. In: Stroebe MS, Hansson RO, Stroebe W, et al., editors. Handbook of Bereavement Research: Consequences, Coping, and Care. Washington, DC: American Psychological Association; 2001

[63] Neimeyer RA. Complicated grief and the quest for meaning: A constructivist contribution. OMEGA Journal of Death and Dying. 2006;**52**(1):37-52. DOI: 10.2190/EQL1-LN3V-KNYR-18TF

[64] Onrust S, Cuijpers P, Smit F, Bohlmeijer E. Predictors of psychological adjustment after bereavement. International Psychogeriatrics. 2007;**19**(5):921-934

[65] Ball K, Lee C. Relationships between psychological stress, coping and disordered eating: A review. Psychology and Health. 2000;**14**(6):1007-1035

[66] Fitzpatrick TR, Tran TV. Bereavement and health among different race and age groups. Journal of Gerontological Social Work. 2002;**37**(2):77-92

[67] Zisook S, Shear K. Grief and bereavement: What psychiatrists need to know. World Psychiatry. 2009;**8**(2):67-74

[68] Grimby A. Bereavement among elderly people: Grief reactions,

post-bereavement hallucinations and quality of life. Acta Psychiatrica Scandinavica. 1993;**87**(1):72-80

[69] Kaprio J, Koskenvuo M, Rita H. Mortality after bereavement: A prospective study of 95,647 widowed persons. American Journal of Public Health. 1987;**77**(3):283-287

[70] Bonanno GA, Wortman CB, Lehman DR, Tweed RG, Haring M, Sonnega J, et al. Resilience to loss and chronic grief: A prospective study from preloss to 18-months postloss. Journal of Personality and Social Psychology. 2002;**83**(5):1150-1164

[71] Bengtson VL, Dowd JJ, Smith DH, Inkeles A. Modernization, modernity and attitudes toward aging: A cross-cultural survey. Journal of Gerontology. 1975;**30**:689-695

[72] Palmore EB, Manton K. Modernization and status of the aged: International correlations. Journal of Gerontology. 1974;**29**:205-210

[73] Cowgill DO. A theory of aging in cross-cultural perspective. In: Cowgill DO, Holmes LD, editors. Ageing and Modernization. New York: Appleton-Century-Crofts; 1972

[74] Cowgill DO. Aging around the World. Belmont, CA: Wadsworth Publishers; 1986

[75] Quadagno J. Aging in Early Industrial Society: Work, Family and Social Policy in Nineteenth Century England. New York: Academic Press; 1982

[76] Giles H, Noels KA, Williams A, Ota H, Lim T-S, Ng SH, et al. Intergenerational communication across cultures: Young people's perceptions of conversations with family elders, non-family elders and same-age peers. Journal of Cross-Cultural Gerontology. 2003;**18**:1-32

[77] Sung K. Elder respect: Exploration of ideals and forms in East Asia. Journal of Aging Studies. 2001;**15**:13-26

[78] Kitayama S, Markus HR, Kurokawa M. Culture, emotion, and well-being: Good feelings in Japan and the United States. Cognition and Emotion. 2000;**14**(1):93-124

[79] Wolfe N. Cross-cultural neuropsychology of aging and dementia: An update. In: Tabor Connor L, Obler LK, editors. Neurobehavior of Language and Cognition. Boston, MA: Springer; 2002

[80] Boelens JJ. Breastfeeding, atopy, and asthma. Lancet. 2003;**361**(9352):174-175

[81] Buysse DJ, Reynolds CF 3rd, Monk TH, Berman SR, Kupfer DJ. The Pittsburgh Sleep Quality Index: A new instrument for psychiatric practice and research. Psychiatry Research. 1989;**28**(2):193-213

[82] De Lau LM, Breteler MM. Epidemiology of Parkinson's disease. Lancet Neurology. 2006;**5**(6):525-535

[83] Federal Interagency Forum on Age-Related Statistics. Older Americans 2008: Key Indicators of Well Being. 2008. Available from: https://agingstats.gov/docs/LatestReport/Older-Americans-2016-Key-Indicators-of-WellBeing.pdf

[84] Hofman A, Brusselle GG, Darwish Murad S, van Duijn CM, Franco OH, Goedegebure A, et al. The Rotterdam study: 2016 objectives and design update. European Journal of Epidemiology. 2015;**30**(8):661-708

[85] Luppa M, Sikorski C, Luck T, Ehreke L, Konnopka A, Wiese B, et al.

Age- and gender-specific prevalence of depression in latest-life: Systematic review and meta-analysis. Journal of Affective Disorders. 2012;**136**(3):212-221

[86] Moher D, Liberati A, Tetzlaff J, Altman DG, The PG. Preferred reporting items for systematic reviews and meta-analyses: The PRISMA statement. PLoS Medicine. 2009;**6**(7):e1000097

[87] Rozenzweig A, Prigerson H, Miller MD, Reynolds CF. Bereavement and late-life depression: Grief and its complications in the elderly. Annual Review of Medicine. 1997;**48**:421-428

[88] Stroup DF, Berlin JA, Morton SC, Olkin I, Williamson GD, Rennie D, et al. Meta-analysis of observational studies in epidemiology: A proposal for reporting. Meta-analysis Of Observational Studies in Epidemiology (MOOSE) group. JAMA. 2000;**283**(15):2008-2012

[89] Tsai WI, Prigerson HG, Li CY, Chou WC, Kuo SC, Tang ST. Longitudinal changes and predictors of prolonged grief for bereaved family caregivers over the first 2 years after the terminally ill cancer patient's death. Palliative Medicine. 2016;**30**(5):495-503

[90] Tsuboya T, Aida J, Hikichi H, Subramanian SV, Kondo K, Osaka K, et al. Predictors of depressive symptoms following the Great East Japan earthquake: A prospective study. Social Science & Medicine. 2016;**161**:47-54

Section 3

Treatment of Senescence

Chapter 6

Anti-Senescence Therapy

Raghad Alshadidi

Abstract

The development of therapeutic strategies aimed at the aging process of cells has attracted increasing attention in recent decades due to the involvement of this process in the development of many chronic and age-related diseases. Interestingly, preclinical studies have shown the success of a number of anti-aging approaches in the treatment of a range of chronic diseases. These approaches are directed against aging processes such as oxidative stress, telomerase shortening, inflammation, and deficient autophagy. Many strategies has been shown to be effective in delaying aging, including antiaging strategies based on establishing healthy lifestyle habits and pharmacological interventions aimed at disrupting senescent cells and senescent-associated secretory phenotype. Caloric restriction and intermittent fasting were reported to activate autophagy and reduce inflammation. In turn, immune-based strategies, senolytic agents, and senomorphics mediate their effects either by eliminating senescent cells through inducing apoptosis or by disrupting pathways by which senescent cells mediate their detrimental effects. In addition, given the association of the decline in the regenerative potential of stem cells with aging, many experimental and clinical studies indicate the effectiveness of stem cell transplantation in preventing or slowing the progress of age-related diseases by enhancing the repairing mechanisms and the secretion of many growth factors and cytokines.

Keywords: age-related disease, senescent-associated secretory phenotype, caloric restriction, senolytic agents, senomorphics, immune-based anti-senescence therapies

1. Introduction

Evidence of the involvement of cellular aging in the development of many age-related diseases, combined with the longevity benefits obtained from preventing the accumulation of senescent cells, raises the possibility that therapeutic targeting of senescent cells extends life span, improves overall health, and delays or prevents the development of age-related diseases. The investigation of the molecular mechanisms accounting for the development of senescent-associated secretory phenotype (SASP) and the ones providing senescent cells maintenance supplied insights for the development of mechanisms to target senescent cells. So far, many approaches have been proposed to target cell senescence, either by inducing the death of senescent cells or by blocking the SASP (**Figure 1**). This chapter provides an overview of senescent cells as an opportunity to intervene in the aging process and presents the various therapeutic anti-senescence paradigms in terms of their molecular mechanisms of action, efficacy, and safety.

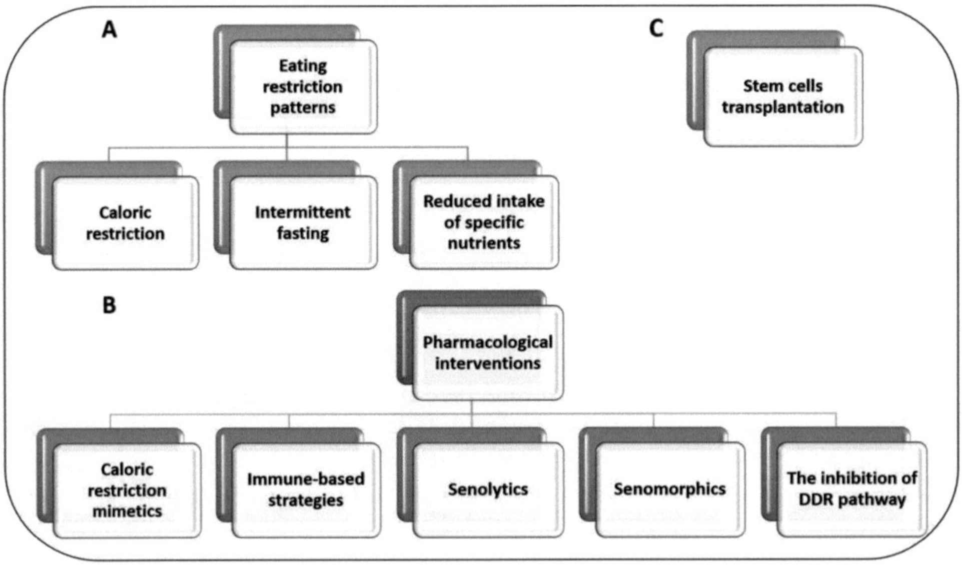

Figure 1.
Strategies targeting cellular senescence. (A) A number of dietary regimens, including caloric restriction, intermittent fasting, and reduced intake of certain nutrients, exert antiaging effects. (B) Pharmacological interventions have been developed to target senescent cells and limit their deleterious effects; certain natural or pharmacological compounds have been reported to exert the beneficial effects of CR, senolytic agents selectively induce apoptosis in senescent cells and block the prosurvival pathway, senomorphic agents interact with the components of the SASP, affecting their upstream pathway or their effectors, immune-based strategies aim to enhance the body's natural defense mechanisms or develop strategies to direct immune cells specifically toward senescent cells and/or overcome the mechanisms that senescent cells use to escape the immune system, and strategies targeting the DDR pathway aim to reduce the induction of cellular senescence. (C) Stem cell transplantation is thought to compensate for the decline in stem cell function, which is partly due to the senescence of stem cells. DDR, DNA damage response. (DDR is the abbreviation of DNA damage response).

2. Eating restriction patterns

A number of dietary regimens have been reported to prolong health span and longevity, in part by reducing their effects on senescence. These regimens include caloric restriction, intermittent fasting, and reduced intake of certain nutrients.

2.1 Caloric restriction (CR)

The underlying premise of caloric restriction (CR) is to reduce calorie availability by ~20–50%, while not consuming fewer vitamins, minerals, and other components of a healthy diet. It is well established that CR is a powerful intervention to extend the average and/or maximum life span of various species including yeast, flies, worms, fish, rodents, and rhesus monkeys [1], improve general health, and decrease aging-associated diseases [1]. Furthermore, data from natural and controlled investigations suggested the beneficial effects of CR on human longevity.

2.1.1 Mechanisms underlying CR antiaging effect

CR induces adaptations in the immune, neuroendocrine, and metabolic system by affecting a number of intracellular pathways; although human and animal observations and studies suggest that CR expands the life span, the mechanism underlying

CR responses has not yet been established, and the followings explain some of the mechanisms, which mediate CR antiaging effects.

CR was reported to decrease cell senescence [2]. Given the fact that cell damage is the main inducer of senescence [3], the cytoprotective properties of CR may explain its senescence rate modulation effects. These effects include decrease in cellular stress [4], decrease in inflammation [5], and increased clearance of damaged proteins and organelles through the activation of autophagy [3]. In addition, CR decreases the mammalian target of rapamycin (mTOR) activity, which, in turn, plays a major role in the activation of cellular senescence [6]. Modulation of nutrient uptake pathways, including insulin-like growth factor (IGF), insulin, mTOR, and AMP-activated protein kinase (AMPK), mainly explains CR-mediated effects [6]. It was reported that CR decreases IGF signaling by inducing hypoglycemia and decreasing the level of insulin [6]. CR-mediated decrease in oxidative damage may be attributed to several mechanisms, including the increase in nitric oxide (NO) concentration [7], the increase in superoxide dismutase (SOD) activity [7], the decrease in reactive oxygen species (ROS) production [3], the decrease in protein glycation [3], the decrease in inflammatory proteins [3], and the increase in the expression of chaperone proteins [3]. Moreover, CR upregulates the expression of sirtuin-2 (sirt-2) [8], and NAD + -dependent protein deacetylases with antioxidant activity [9], the existence of extra copies of which was shown to extend the life span by up to 30%. CR exerts anti-inflammatory effect; a large number of investigations have shown that it reversed the effects of SASP and modulate age-related chronic inflammatory conditions [5]. This anti-inflammatory response can be attributed to the regulation of the activity of pro-inflammatory upstream signaling pathway molecules such as MAPKs (ERK, JNK, and p38), and NIK/IKKs and the suppression of key pro-inflammatory mediators such as NF-B, IL-1, IL-6, TNF, cyclooxygenase 2 (COX-2), and inducible nitric oxide synthase (iNOS) [5]. Intriguingly, CR was reported to induce a slight increase in circulating cortisol, which also account for the reduction in systemic inflammation [3]. Additionally, CR was reported to enhance DNA repair mechanism. This effect is mediated by the decrease of age-dependent decline in non-homologous end joining (NHEJ), the decrease of the age-dependent decline of polymerase alpha and beta and the increase of their fidelity, the induction of the base excision repair pathway (BER), and the enhancement of nucleotide excision repair (NER) [3]. Moreover, CR was shown to alter the phenotypes of stem cells, improve their function, and promote their self-renewal in mice [10]. Although most studies proposed the beneficial effects of CR, it is noteworthy to report the results of one study that suggest the negative impact CR has on brain integrity of mouse lemurs without affecting cognitive performances [11]. More investigations are required to evaluate the long-term effects of CR and create a comprehensive full image of the molecular and cellular mechanisms underlying its antiaging effect.

2.2 Other type of dietary restriction with antiaging effects

In addition to CR, other approaches have been proposed to reduce the effects of aging, including intermittent fasting and the reduced intake of certain nutrients. Intermittent fasting (IF) is an eating pattern that switches between fasting and eating on a regular schedule. It has been shown that IF has a protective effect against many age-related diseases such as obesity, hyperinsulinemia, hepatic steatosis, and inflammation [12]. Many types of IF have been proposed, including a fasting-mimicking diet, which has been reported to prolong the life span, reduce visceral fat, reduce cancer and skin damage, rejuvenate the immune system, and slow bone mineral loss in mice [13]. Other studies have shown that reduced intake of specific nutrients exerts antiaging effects. For example, reducing protein intake has been reported to reduce the risk of cancer death and overall mortality [12]. Interestingly, it was suggested that the life span benefits of dietary restriction can be obtained from the reduced intake of certain amino acids such as tryptophan and methionine [14]. In addition, ketogenic diet, which primarily consists of high fats, moderate proteins, and very low carbohydrates, was widely reported to extend longevity and health span in mice [15].

3. Pharmacological interventions

The involvement of cell senescence in aging and the development of many age-related diseases has stimulated efforts to develop a number of strategies aimed at eliminating senescent cells and/or limiting their deleterious effects. These strategies include CR mimetics, senolytic agents, senomorphic agents, immune-based strategies, and strategies targeting the DNA damage response (DDR) pathway.

3.1 CR mimetics

CR mimetics are agents that have the beneficial effects of CR without the need to struggle with diet limitation; the followings are some of the most well-known CR mimetics.

3.1.1 Resveratrol

Resveratrol (3,5,4′-trihydroxystilbene) is a natural plant-derived polyphenolic, phytoalexin compound found in grapes, cranberries, and peanuts. Resveratrol has been long used in traditional medicine, and now, it has wide range of applications in modern medicine thanks to their antioxidant, anti-inflammatory, anti-obesity, anti-diabetic, antibacterial, anticarcinogenic, cardioprotective, and immunomodulating properties [16]. It is suggested that resveratrol exerts its antiaging effects through the activation of sirtuin-1 (sirt-1) [12]. One clinical study on healthy, obese men reported that 30 days of resveratrol supplementation elevated intramyocellular lipid levels, and decreased intrahepatic lipid content, circulating glucose, triglycerides, alanine-aminotransferase, inflammation markers, and systolic blood pressure with an improvement in HOMA index [17]. Another clinical study on overweight older individuals provides evidence that supplementation of resveratrol improves memory performance in association with improved glucose metabolism and increased hippocampal functional connectivity in older adults [3].

3.1.2 Metformin

Metformin is an FDA-approved antidiabetic agent used as a first-line drug for treating type 2 diabetes mellitus. The antiaging effects of metformin are well established in *Caenorhabditis elegans*, *Drosophila melanogaster*, and mice [5]. Metformin was reported to exert its antiaging effects through the activation of AMPK, which, in turns, leads to the inhibition of mTOR [18]. The wide application, well-known pharmacokinetics and acceptable toxicity encourage the use of metformin as an antiaging drug [18]. However, further investigations are required since the metformin's antiaging effect has not been identified in humans yet [5].

3.1.3 Rapamycin

Rapamycin is an FDA-approved immunosuppressant, which is extensively used following kidney and liver transplants and to treat certain types of cancers and complications of tuberous sclerosis [19]. The antiaging effect of rapamycin is well established for several years in model organisms including mice [19] and is mediated through the direct inhibition of the kinase activity of mTOR [5]. Despite its widely reported antiaging effect, certain side effects pose a barrier to the use of rapamycin as antiaging agent; this includes hyperlipidemia, hypercholesterolemia, and hypertriglyceridemia, glucose intolerance, insulin resistance and new-onset diabetes, anemia and thrombocytopenia, dermatological events, gastrointestinal disorders, sinusitis, respiratory and urinary infections, and testicular dysfunction [19]. To minimize these side effects, a number of alternative treatment regimens have been developed, including intermittent rapamycin [19]. In addition, fewer side effects have been reported after the administration of rapamycin analogs with a reduced effect on glucose metabolism, such as everolimus and temsirolimus [19].

3.2 Enhancement of the immune clearance of senescent cells

The immune system has its own internal targeting mechanism for senescent cells. Multiple components of the immune system target senescent cells, including NK cells, T cells, and macrophages [20]. The immunogenicity of senescent cells, combined with the age-related decline in immune function, raises the potential for strategies such as boosting the immune system or targeting the inhibitory mechanisms by which senescent cells escape the immune system to be exploited to enhance the clearance of senescent cells [20]. In this context, many strategies have been proposed, including blocking the inhibitory decoy receptor 2 (DR2) and stimulating the innate immune response using the viral infection stimulator poly (I:C) [21]. In addition, the implementation of the advances in genetic engineering by designing chimeric antigen receptor T (CAR T) cells specific for senescent cells is a promising approach. However, the lack of overlap between the extracellular markers identified by different studies is an obstacle for designing CAR T cells specific for senescent cells [20].

3.3 Senolytic agents

Senolytic agents target specifically senescent cells, and they are considered promising agents for delaying aging processes as they target the fundamental mechanisms that are contributors for many diseases. Dasatinib, quercetin, and fisetin are the most well-studied senolytic agents [22]. Dasatinib is a second-generation tyrosine kinase inhibitor that is used for the treatment of CML and AML [23]. Quercetin is a polyphenolic flavonoid compound with antioxidant properties, which exerts preventive effects for various diseases, such as osteoporosis, some forms of cancer, tumors, and lung and cardiovascular diseases [24]. Fisetin is a flavonol that shows potential as an anti-inflammatory, chemopreventive, and chemotherapeutic agent [25]. Senolytic agents were discovered by scanning using bioinformatic approach to find drugs that disrupt the senescent cell anti-apoptotic pathways (SCAPs) network nodes, which differ from the one-target one-drug approach. Thereby, an important characteristic of these agents is their targeting to multiple SCAP network nodes rather than acting upon single or limited targets, which, in turn, reduces the off-target apoptotic effects on nonsenescent cell types [22]. *In vivo* studies showed that these agents reduce senescent cells by apoptosis; specifically, the underlying molecular mechanisms by which senolytic agents mediate their effects are lowering of p16Ink4a, targeting Bcl-2 family, hypoxia-inducible factor 1-alpha (HIF-1a), and other SCAPs network components [22]. Since dasatinib and quercetin are senolytics for different cell lines, it is suggested that the effects of senolytic drugs depend on the type of senescent cells [22]. Interestingly, the combination of dasatinib + quercetin was reported to be senolytic for cell lines in which neither dasatinib nor quercetin are senolytics on their own [22]. Senolytic drugs were reported to improve cardiac function in mice, enhance insulin sensitivity, reduce the adipose tissue inflammation, and alleviate many age-related diseases in which the accumulation of senescent cells plays role in its pathogenesis including Alzheimer's disease, chronic lung diseases, osteoporosis, and intervertebral disk disease [22]. As senolytic drugs are new agents, special cautions are considered while testing them in clinical trials; therefore, they entered only clinical trials for serious diseases that lack effective treatment strategies [22]. The results of a clinical trial where dasatinib + quercetin were administered orally by patients with diabetes complicated by renal dysfunction showed decrease in senescent cells and adipose tissue inflammation [22]. Many clinical trials are now ongoing or planned including trials of dasatinib + quercetin for the treatment of many senescence and age-related diseases [22]. So far, senolytic agents are not used out of clinical trials as their effects are under investigation [22]. However, depending on the results of clinical trials, senolytic agents can have further applications in the future to delay and prevent the developing of senescence- and age-related diseases in those subjects, in whom the presence of senescent cells can be detected in body fluids or by imaging [22].

3.4 Senomorphic agents

Senomorphic agents are an alternative approach to senolytics, the concept behind this approach is to disrupt pathways by which senescent cells mediate its detrimental effects without eliminating the cells. For this purpose, neutralizing antibodies targeting SASP components or their receptors have been developed [26]. Approaches based on the transcriptional modulation of the expression of SASP factors were developed to reduce SASP production [27].

In addition, based on the fact that mTOR activation promotes SASP production through translation of subsets of mRNA that stabilizes many cytokine-encoding transcripts, certain mTOR inhibitors can be considered senomorphic agents [20]. For example, rapamycin, which exhibits CR-mimicking effects, was reported to decrease SASP production by inhibiting mTOR [20]. Besides, apigenin and kaempferol were shown to attenuate SASP production through their modulating effects on NF-κB signaling [20].

Senolytic agents have advantages over senomorphics, which include the possibility to take them intermittently and the reduction of the likelihood of senescence bypassing mutations that can promote tumorigenesis, since these agents eliminate senescent cells rather than targeting SASP production [20]. However, the still undetermined safety of prolonged or repeated administration of senolytic agents combined with an emerging study report on the likelihood of senolysis damage to cells with structural functions [28] makes senomorphics a considerable alternative to senolytic agents.

3.5 Targeting senescent cells by inhibiting the DNA damage response (DDR) pathway

DNA damage response (DDR) pathway is a signaling cascade that is activated in response to DNA damage. Given the evidence that activation of the DDR pathway triggers cell senescence, antisense oligonucleotides have been developed to inhibit telomeric DDR. Results from *in vitro* and *in vivo* studies in mice have provided evidence for the effectiveness of this approach [20].

4. Stem cell transplantation

Cell senescence has been implicated in the decline of stem cells' function and proliferation potential [20], which, in turn, contributes to aging and the development of age-related disease [29]. Therefore, stem cell transplantation has been proposed as a strategy to treat many age related disease including Alzheimer's disease [30], macular degeneration [31], osteoarthritis [32], and frailty [33]. The beneficial effects of this strategy are exerted through the compensation of aging-related decline in stem cell function, the regulation of inflammation and immune responses, as well as the secretion of therapeutic cytokines and factors [29]. The promising results of the preclinical experiments on mice models [33] lead to its translation to clinical trials. Phases I and II clinical trials were conducted to investigate the efficiency of mesenchymal stem cells (MSCs) infusion in alleviating frailty and the results showed safety profile and promising therapeutic efficacy [33]. A number of clinical trials provided evidence of the efficiency of MSCs therapies for the treatment of osteoarthritis [32]. Furthermore, a phase I clinical study supports the efficiency and safety of the transplantation of embryonic stem cell-derived retinal pigment epithelium patches as a regenerative strategy for age-related macular degeneration [31]. Although the results of preclinical and clinical studies provide initial evidence of the efficiency of stem cell transplantation for the treatment of age-related diseases, the concerns of stem cells tumorigenicity impose the need for further research and clinical trials with the consideration of the framework regulatory agencies to ensure the safety of participants [29].

5. Conclusion

Despite initial evidence for the safety and efficacy of a number of antiaging therapeutic approaches to combat the aging process, the novelty of this area of study stresses the need to conduct further preclinical and clinical study to understand the efficacy, safety, and long-term effects of anti-senescence therapeutic strategies in addition to the optimization of the most effective strategy with minimal off-targets and side effects. Further research in terms of the molecular mechanism of senescence and anti-senescence therapeutic strategies could reshape our view of health management during aging, offering many therapeutic options in the context of increasing longevity and preventing or alleviating many age-related diseases.

Acronyms and abbreviations

AMPK	AMP-activated protein kinase
CAR-T	cell chimeric antigen receptor (CAR) T cell
COX-2	cyclooxygenase 2
CR	caloric restriction
DDR	DNA damage response
DR2	decoy receptor 2

HOMA index	homeostatic model assessment index
IF	intermittent fasting
IGF 1	insulin-like growth factor 1
IKK	inhibitor of NF-kappaB kinase
iNOS	inducible nitric oxide synthase
MAPK	mitogen-activated protein kinase
mTOR	the mammalian target of rapamycin
NIK	NF-κB-inducing kinase
NO	nitric oxide
SASP	senescent-associated secretory phenotype
SCAP	senescent cell anti-apoptotic pathways
sirt-2	sirtuin
SOD	super oxide dismutase

Author details

Raghad Alshadidi
Kazan Federal University, Kazan, Russia

*Address all correspondence to: rag.rsh997@gmail.com

References

[1] Fontana L, Partridge L, Longo VD. Dietary restriction, growth factors and aging: From yeast to humans. Science. 2010;**328**(5976):321. DOI: 10.1126/SCIENCE.1172539

[2] Fontana L, Nehme J, Demaria M. Caloric restriction and cellular senescence. Mechanisms of Ageing and Development. 2018;**176**:19-23. DOI: 10.1016/j.mad.2018.10.005

[3] Witte AV, Kerti L, Margulies D, Flöel A. Effects of resveratrol on memory performance, hippocampal functional connectivity, and glucose metabolism in healthy older adults. The Journal of Neuroscience. 2014;**34**(23):7862-7870. DOI: 10.1523/JNEUROSCI.0385-14.2014

[4] Barja G. Aging in vertebrates, and the effect of caloric restriction: A mitochondrial free radical production-DNA damage mechanism? Biological Reviews of the Cambridge Philosophical Society. 2004;**79**(2):235-251. DOI: 10.1017/S1464793103006213

[5] Kim DH, Bang E, Jung HJ, Noh SG, Yu BP, Choi YJ, et al. Anti-aging effects of calorie restriction (CR) and CR mimetics based on the senoinflammation concept. Nutrients. 2020;**12**(2):422. DOI: 10.3390/nu12020422

[6] Weichhart T. mTOR as regulator of lifespan, aging and cellular senescence. Gerontology. 2018;**64**(2):127. DOI: 10.1159/000484629

[7] Stankovic M, Mladenovic D, Ninkovic M, Vucevic D, Tomasevic T, Radosavljevic T. Effects of caloric restriction on oxidative stress parameters. General Physiology and Biophysics. 2013;**32**:277-283. DOI: 10.4149/gpb_2013027

[8] Xiang L, He G. Caloric restriction and antiaging effects. Annals of Nutrition & Metabolism. 2011;**58**(1):42-48. DOI: 10.1159/000323748

[9] Singh CK, Chhabra G, Ndiaye MA, Garcia-Peterson LM, Mack NJ, Ahmad N. The role of sirtuins in antioxidant and redox signaling. Antioxidants & Redox Signaling. 2018;**28**(8):643-661. DOI: 10.1089/ARS.2017.7290

[10] Schultz MB, Sinclair DA. When stem cells grow old: Phenotypes and mechanisms of stem cell aging. Development. 2016;**143**(1):3. DOI: 10.1242/DEV.130633

[11] Pifferi F, Terrien J, Marchal J, Dal-Pan A, Djelti F, Hardy I, et al. Caloric restriction increases lifespan but affects brain integrity in grey mouse lemur primates. Communications Biology. 2018;**1**:30. DOI: 10.1038/s42003-018-0024-8

[12] Hanjani NA, Vafa M. Protein restriction, epigenetic diet, intermittent fasting as new approaches for preventing age-associated diseases. International Journal of Preventive Medicine. 2018;**9**:58. DOI: 10.4103/IJPVM.IJPVM_397_16

[13] Brandhorst S et al. A periodic diet that mimics fasting promotes multi-system regeneration, enhanced cognitive performance, and healthspan. Cell Metabolism. 2015;**22**(1):86-99. DOI: 10.1016/J.CMET.2015.05.012

[14] Grandison RC, Piper MDW, Partridge L. Amino-acid imbalance explains extension of lifespan by dietary restriction in *Drosophila*. Nature. 2009;**462**(7276):1061-1064. DOI: 10.1038/nature08619

[15] Roberts M, Wallace M, Tomilov A, Zhou Z, Marcotte G, Tran D, et al. A ketogenic diet extends longevity and healthspan in adult mice. Cell Metabolism. 2017;**26**(3):539-546.e5. DOI: 10.1016/J.CMET.2017.08.005

[16] Gianchecchi E, Fierabracci A. Insights on the effects of resveratrol and some of its derivatives in cancer and autoimmunity: A molecule with a dual activity. Antioxidants. 2020;**9**(2):91. DOI: 10.3390/ANTIOX9020091

[17] Timmers S, Konings E, Bilet L, Houtkooper R, Weijer T, Goossens G, et al. Calorie restriction-like effects of 30 days of Resveratrol (resVida™) supplementation on energy metabolism and metabolic profile in obese humans. Cell Metabolism. 2011;**14**(5):612-622. DOI: 10.1016/J.CMET.2011.10.002

[18] Onken B, Driscoll M. Metformin induces a dietary restriction-like state and the oxidative stress response to extend *C. elegans* healthspan via AMPK, LKB1, and SKN-1. PLoS One. 2010;**5**(1): e8758. DOI: 10.1371/JOURNAL.PONE.0008758

[19] Apelo S, Lamming D. Rapamycin: An inhibitor of aging emerges from the soil of Easter Island. The Journals of Gerontology. Series A, Biological Sciences and Medical Sciences. 2016;**71**(7):841-849. DOI: 10.1093/GERONA/GLW090

[20] Di Micco R, Krizhanovsky V, Baker D. Cellular senescence in ageing: From mechanisms to therapeutic opportunities. Nature Reviews. Molecular Cell Biology. 2021;**22**(2):75-95. DOI: 10.1038/s41580-020-00314-w

[21] Ovadya Y, Krizhanovsky V. Strategies targeting cellular senescence. The Journal of Clinical Investigation. 2018;**128**(4): 1247-1254. DOI: 10.1172/JCI95149

[22] Gerdes EOW, Zhu Y, Tchkonia T, Kirkland JL. Discovery, development, and future application of senolytics: Theories and predictions. The FEBS Journal. 2020;**287**(12):2418. DOI: 10.1111/FEBS.15264

[23] Hořínková J, Šíma M, Slanař O. Pharmacokinetics of Dasatinib. Prague Medical Report. 2019;**120**(2-3):52-63. DOI: 10.14712/23362936.2019.10

[24] Xu D, Hu M-J, Wang Y-Q, Cui Y-L. Antioxidant activities of Quercetin and its complexes for medicinal application. Molecules. 2019;**24**(6):1123. DOI: 10.3390/MOLECULES24061123

[25] Grynkiewicz G, Demchuk OM. New perspectives for fisetin. Frontiers in Chemistry. 2019;7:697. DOI: 10.3389/FCHEM.2019.00697

[26] Liu S, Uppal H, Demaria M, Desprez P-Y, Campisi J, Kapahi P. Simvastatin suppresses breast cancer cell proliferation induced by senescent cells. Scientific Reports. 2015;**5**(1):1-11. DOI: 10.1038/srep17895

[27] Cuollo L, Antonangeli F, Santoni A, Soriani A. The senescence-associated secretory phenotype (SASP) in the challenging future of cancer therapy and age-related diseases. Biology. 2020;**9**(12):1-16. DOI: 10.3390/BIOLOGY9120485

[28] Grosse L, Wagner N, Emelyanov A, Molina C, Lacas-Gervais S, Wagner K, et al. Defined p16 high senescent cell types are indispensable for mouse healthspan. Cell Metabolism. 2020;**32**(1):87-99.e6. DOI: 10.1016/J.CMET.2020.05.002

[29] Nguyen H, Zarriello S, Coats A, Nelson C, Kingsbury C, Gorsky A, et al. Stem cell therapy for neurological disorders: A focus on aging.

Neurobiology of Disease. 2019;**126**:85. DOI: 10.1016/J.NBD.2018.09.011

[30] Vasic V, Barth K, Schmidt MHH. Neurodegeneration and neuro-regeneration Alzheimer's disease and stem cell therapy. International Journal of Molecular Sciences. 2019;**20**(17):4272. DOI: 10.3390/IJMS20174272

[31] da Cruz L et al. Phase 1 clinical study of an embryonic stem cell-derived retinal pigment epithelium patch in age-related macular degeneration. Nature Biotechnology. 2018;**36**(4): 328-337. DOI: 10.1038/NBT.4114

[32] Freitag J, Bates D, Boyd R, Shah K, Barnard A, Huguenin L, et al. Mesenchymal stem cell therapy in the treatment of osteoarthritis: Reparative pathways, safety and efficacy—A review. BMC Musculoskeletal Disorders. 2016;**17**(1):230. DOI: 10.1186/S12891-016-1085-9

[33] Schulman I, Balkan W, Hare J. Mesenchymal stem cell therapy for aging frailty. Frontiers in Nutrition. 2018;**5**:108. DOI: 10.3389/FNUT.2018.00108